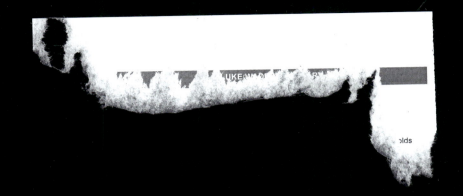

Multimedia Management

INNOVATIVE TECHNOLOGY SERIES
INFORMATION SYSTEMS AND NETWORKS

Multimedia
Management

edited by
José Neuman de Souza
& Nazim Agoulmine

HPS
HERMES PENTON SCIENCE

First published in 2000 by Hermes Science Publications, Paris
First published in 2002 by Hermes Penton Ltd
Derived from Networking and Information Systems Journal, *Multimedia Management*, Vol. 3, no. 2.

Hermes Penton Science
120 Pentonville Road
London N1 9JN

British Library Cataloguing in Publication Data

A CIP record for this book is available from the British Library.

ISBN 1 9039 9623 6

Typeset by Saxon Graphics Ltd, Derby
Printed and bound in Great Britain by Biddles Ltd, Guildford and King's Lynn
www.biddles.co.uk

Contents

Foreword

For years, industry has predicted and anticipated the arrival of multimedia services via the network. From all indications, this new age is upon us now especially with the emergence of the Internet and the huge potential of e-commerce. The study of multimedia transmission over various network technologies has been the subject of research by research teams all over the world in industrial and research laboratories. Much of that work has focused on QoS (Quality of Service) management in B-ISDN (broadband integrated services digital) networks such as ATM, and recently, IP-based networks. Whilst originally these two technologies were considered as antagonist, the actual enhancement of operator's core networks show, on the contrary, an integrated approach to the core network (with MPLS) and the access network (with ADSL). The integrations of various technologies using sophisticated load balancing and routing techniques will allow the deployment of new intelligent approaches to control the traffic and congestion at different levels of the network to provide the end user with QoS customised multimedia communications.

The contributions within this publication are based on presentations made at the Networking 2000 main conference, held in Paris, France, on May 19, 2000. These include:

Collaborative Virtual Environments: Managing the Shared Spaces. This paper presents a novel approach to collaboration based on the concept of shared space. This approach was proposed for use in the development of new multimedia collaborative applications.

QoS Management for Multimedia Applications Using an Efficient Queue Policy. This paper presents state of art of queue policies for control of parameters and a novel approach adapted to the control of multimedia applications streams.

Structuring Devolved Responsibilities in Network and Systems Management. This paper presents the concept of policy management. In this approach, the management of network QoS is designed to use policy models that define the entire enterprise rights of resource usage. A formal language was specified to facilitate the definition of such a model.

An Agent-based Framework for Large-scale Internet Applications. This paper presents a new approach to service discovery and QoS negotiation over the Internet.

In it is proposed a framework for service discovery and QoS negotiation at the network level that relies on two concepts: multi-agent systems and agent communication languages (ACL).

An Agent-based Paradigm for Managing Service Quality in Open Network Environments. This paper presents an agent-based approach to management of QoS in IP-based networks. This approach is based on the concept of Active Networks and Mobile Agents. The papers show how an agent can collaborate to provide the user with the best cost-effective service in a multi-operator environment.

Adaptive Multicast Group Management for Distributed Event Correlation. This paper presents a novel approach to management. The increase in enterprise network size makes their management very onerous and time consuming. This paper proposes a new approach to the management of network elements as a whole, using group communications between management agents in order to reduce the management traffic and increase the management process efficiency.

José Neuman de Souza
Federal University of Ceará, Brazil

Nazim Agoulmine
University of Evry, France

Chapter 1

Collaborative virtual environments: managing the shared spaces[1]

Hui Zhao and Nicolas D. Georganas

Multimedia Communication Research Laboratory, School of Information Technology and Engineering, University of Ottawa, Canada

1. Introduction

1.1. Distributed interactive simulation to collaborative virtual environments

The venerable problem-solving technique using simulation finds itself in the midst of a revolution. As an economical, timesaving and safety solution, simulation is widely applied today to support a myriad of purposes, including: training, interaction, visualization, hardware testing, and real-time decision support.

Distributed interactive simulation (DIS) is a newer concept, accompanying the development of computer network technology. DIS refers to the technology of executing simulation entities over multiple computers connected via a network through which simulation entities' messages are exchanged.

DIS adds advanced features to simulation, including: reduced model execution time; scalable performance; geographically distributed users and/or resources; integrated simulations running on different platforms; fault tolerance.

Due to the extraordinary development of computer networks, distributed interactive simulation (DIS) plays a more and more important role in the simulation community. In general, there are three categories of DIS applications. The first category is pure simulation, which means that all simulation entities are computer programs, for example, weather forecast simulation, celestial mechanics simulation, etc. The second category is the people-in-the-loop simulation when the human acting as user or trainee closes the control loop. This is widely used in virtual shopping malls, collaborative design and engineering, collaborative augmented reality for sharing space, multi-user virtual conferencing, shared virtual environment for training (pilot training), network interactive 3-D games,

---

[1] This work was supported in part by a research contract from the Communications Research Centre (CRC) of Industry, Canada and in part by Communications and Information Technology, Ontario, a Centre of Excellence.

etc[5]. The third category is equipment-in-the-loop simulation in which at least one actual equipment is connected to the simulation network and can interact with other computer program simulation entities. Example applications include aeroplane design and testing, weapon systems design and testing, etc.

As a branch of DIS, the real-time distributed interactive simulation (RT-DIS) refers to a DIS with special time requirements. The term "real-time", as it relates to simulation, requires that the computer program execution of a modeled, dynamic process must occur in the real-world time (i.e., not faster or slower)[6]. It represents or simulates a real, dynamic phenomenon as it occurs. Real-time simulation allows a more realistic representation of the physical system being studied (as opposed to pure mathematical analysis or standard computer analysis) and permits both quantitative and qualitative (human analysis) evaluation[6]. Obviously, all people-in-the-loop simulations and many equipment-in-the-loop simulations are some kind of RT-DIS application with a different time critical level.

As a specific application of RT-DIS, a collaborative virtual environment (CVE) is a software system in which multiple users interact with each other in real-time, even though those users may be physically located in different places around the world. The CVE enabled solution will affect remarkably many applications that demand a high quality human machine interface (HMI), such as e-commerce, network entertainment, distance learning, etc. The CVE systems usually aim at providing users with a sense of realism by incorporating realistic 3-D graphics and vivid sound to create an immersive experience.

1.2. State of the art in advanced simulation techology and standards

1.2.1. SIMNET

In the mid-1980s the U.S. Department of Defence, Defence Advanced Research Projects Agency (DARPA), launched the SIMNET project and generated a real-time vehicle-level distributed interactive (virtual) simulation system. In SIMNET[7], individual vehicle simulators are connected via a computer network, permitting them to coexist in a common, shared simulation environment and to interact (e.g. engage in combat) through the exchange of information packets on the network that connects them. SIMNET simulators usually each represent a single tank or vehicle. SIMNET is used to train tank and vehicle crews in cooperative team tactics.

SIMNET follows four design principles: (1) distributed computation based on combat entity ownership, (2) avoidance of single critical resources, (3) reliance on broadcast communications, and (4) replication of a limited set of combat entity attributes among all simulations.

SIMNET was the first step for distributed interactive simulation.

1.2.2. Aggregate Level Simulation Protocol (ALSP)

In early 1990, DARPA sponsored MITRE to investigate the design of a general interface between large, existing, aggregate-level combat simulations. MITRE presented the ALSP[8] system and adopted all the principles from SIMNET. In addition, aggregate-level simulations have unique requirements for time and data management that are addressed by services specific to ALSP. ALSP provides time management services to coordinate simulation times and preserve event causality across simulations. The data management scheme allows each simulation to use its own representation of data. Simulations share information in a commonly understood manner independent of each simulation's internal data representation. The typical application based on ALSP is the Joint Training Confederation (JTC)[8].

The main difficulty of ALSP has been its inflexibility. In order to incorporate a new simulation in the federation, considerable re-writing of the existing members is needed.

1.2.3. Distributed Interactive Simulation (DIS)

The DIS protocol is intended to replace its precursor – the SIMNET protocol. DIS had larger ambitions than SIMNET. SIMNET devices used the same technology, whereas DIS was aimed at simulators using dissimilar technology[9]. In 1993, DIS was approved as the IEEE 1278 series standards.

Through the use of the DIS protocol standard, DIS integrates traditional simulator technologies with computer communication technologies to create a system that provides a common field on which the various simulators can interact in active, real-time situations.

DIS was founded on a few basic concepts: multiple entities simulation; no central node; autonomous simulation nodes; standard communications protocol; receiving nodes perception; local maintenance of other nodes' states.

Although it was a great improvement on SIMNET, several major problems associated with scaling the current suite of DIS protocols illustrate the difficulty of building large virtual environments. First, enormous bandwidth and computational requirements were required for large-scale simulation; second, multiplexing of different media at the application layer was needed; third, DIS lacked an efficient method for handling static objects; lastly, models and world databases must be replicated at each simulator.

1.2.4. Advanced Distributed Simulation (ADS)

There are areas of simulation where DIS may not be appropriate or meet the timing or data transmission rates required. Founded on DIS, ADS[10] uses similar principles that DIS is founded upon, but allows for use of protocols and methodologies outside of the DIS standards[11]. The term ADS includes DIS as a subset and is intended to support a mixture of virtual, live, and constructive entities.

ADS is more conceptual and therefore more flexible than DIS. DIS can be thought of as one specific implementation of the ADS concept. The High Level Architecture (HLA) is another example of an implementation of the ADS concept[11].

1.2.5. Scalable Platform for Large Interactive Networked Environments (SPLINE)

To support their work on social virtual reality, Mitsubishi Electric Research Laboratories (MERL) researchers designed and implemented the SPLINE middleware system with many features[12]: multiple users; spoken interaction; 3-D graphics and sound; run-time modifiability to the environment; open interfaces at both the network and application programmer levels.

SPLINE provides development APIs and libraries. Such libraries provide very detailed and essential services for real-time multi-user cooperative applications. For its communication, SPLINE uses the Interactive Sharing Transfer Protocol (ISTP), which is a hybrid protocol supporting many modes of transportation for VR data and information through five sub-protocols, namely: 1–1 Connection; Object State Transmission; Streaming Audio; Locale-Based Communications and Content-Based Communication sub-protocols.

SPLINE partitions the World Model in *locales* which may have any shape. Once a user joins a given *locale* everything which is located in that *locale* as well as the *locales* with immediate neighbourhood would be visible. It is possible to be "present" in more than one *locale* by using spObserver objects. Internally, a spObserver object makes the local SPLINE engine listen to the multicast group of that *locale* (and its neighbourhood).

SPLINE has the advantage of providing audio communication and audio/visual rendering modules. But since it is derived from SIMNET and DIS, the major problems of the DIS system also exist in SPLINE

1.2.6. High Level Architecture (HLA)

HLA is the next generation DIS adopting new architecture and bandwidth reduction techniques. HLA is inspired by all previous architectures and protocols and designed to supplant both DIS and ALSP[13].

HLA was proposed by DARPA/DMSO for distributed simulation in 1996. It represents an attempt to build on ALSP and DIS and to apply the lessons learned to provide a much more permanent foundation and framework for distributed simulation in the future. Every simulation developed for the US Department of Defense will be expected to conform to HLA from October 1, 2000. The Object Management Group (OMG) has adopted HLA as the Facility for Distributed Simulation Systems 1.0 in November 1998[3]. More recently (November 2000), HLA was approved as the IEEE 1516 series standard for distributed interactive simulation[4]. Obviously, the HLA will be the infrastructure for future distributed simulation.

2. High Level Architecture

2.1. High Level Architecture standards

The HLA standard for modeling and simulation (M&S) consists of three specifications:

- IEEE1516 – Framework and Rules[14]
  This is the top level base document of a family of related High Level Architecture (HLA) documents. It defines the HLA, its components, and the responsibilities of federates and federations.
- IEEE1516.1 – Federate Interface Specification[15]
  Defines the functional interface between federates and the HLA run time infrastructure (RTI), including six service groups: federation management, declaration management, object management, ownership management, time management and data distribution management.
- IEEE1516.2 – Object Model Template (OMT) Specification[16]
  Provides a specification for documenting key information about simulations and federations. OMT is used to describe Simulation and Federation Object Models (SOMs and FOMs).

Interoperability and reusability are major drivers for the HLA. Interoperability is the ability of system components to exchange data and the ability of those components to interpret the data in a consistent way. Reusability is facilitated by having modular components with commonly understood behaviours and well-defined interfaces through which a variety of client applications can access the components.

2.2. Architecture and mechanisms

Federation and federates are two important concepts in HLA. Federation is a named set of interacting federates that are used as a whole to achieve some specific objective. A *federate* is a member of a HLA federation that provides specific function to other federates. All applications participating in a *federation* are called federates.

The information is exchanged among federates through a kind of group communication mechanism. Data source federates publish a variable to a federation and data consumer federates subscribe to the same variable. For each variable, there is a group containing all data consumer federates. When the source federate updates the variable, all the consumer federates will be notified and get the updated value of the variable. The variable update is implemented using a multicast mechanism.

Run-time Infrastructure (RTI) is a software that implements the interface specification of the HLA standard. It provides services in a manner that is analogous to the way a distributed operating system provides services to applications[17]. RTI provides a C++ library, libRTI, through which a federate developer can use the services specified in the HLA Interface Specification, refer to Figure 1.

Within libRTI, the class *RTIAmbassador* bundles the services provided by the RTI. All requests made by a federate on the RTI take the form of a *RTIAmbassador* method call. The abstract class *FederateAmbassador* identifies the callback functions each federate is obliged to provide.

RTI and Federate "Ambassadors"

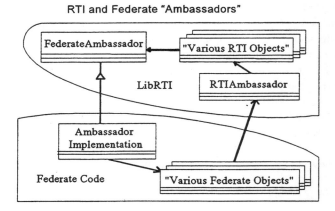

Figure 1. *Federate and libRTI*[17]

More than twenty companies and institutes are manufacturing and supporting RTI and OMT software. A vendors list could be found at DMSO HLA homepage[18].

2.3. HLA applications

The objective for military application of HLA is to construct a rapidly configured mix of computer simulations, actual war-fighting systems, and weapons systems simulators geographically distributed and networked, involving tens of thousands of entities to support training, analysis, and acquisition. Such simulations would be used both to train individuals to perform particular tasks, to interpret data, and to make decisions, and to help groups of individuals (tank crews, fighter squadrons) work together as a team.

Besides newly developed HLA-based simulations, all legacy military simulations are being ported to HLA since every simulation developed for the US Department of Defense will be expected to conform to HLA from October 1, 2000.

For example, the U.S. Army's Tank-Automotive Research Development and Engineering Center's (TARDEC) VETRONICS Simulation Facility (VSF) and Physical Simulation Laboratory (PSL) are integrating man-in-the-loop distributed simulations with two six-degrees of freedom motion-based systems[19]; refer to Figure 2. The VSF is capable of networking with other simulators via Distributed Interactive Simulation (DIS) protocol. They declared that the VSF would migrate from a DIS-based simulation facility to a fully compliant High Level Architecture (HLA) facility[19].

Civil industry would also benefit from HLA for many applications. Roger Smith, the technical director of BTG inc, USA,[20] emphasised in the IEEE DSRT2000 workshop[21] that the distributed simulation would go beyond military

Figure 2. TARDEC's VSF and PSL

applications. It will be widely used in many civil fields, including: massively multi-user games, distributed theme parks, commercial distributed computing, weather analysis and prediction, Internet traffic analysis, stock market prediction, real-time control, traffic/air traffic control, distributed engineering, movie production, virtual conference, e-commerce, etc.

The virtual shopping mall[22,23] developed at the Multimedia Communication Research Labortary (MCRLab), the University of Ottawa, is a good example; refer to Figure 3[23]. With the creation of a virtual shopping mall, simulations of most of the actual shopping environments and user interactions can be achieved. The

Figure 3. MCRLab's virtual shopping mall[23]

virtual mall brings together the services and inventories of various vendors and provides customers with the same shopping experience as they would have in an actual store or shopping mall. Real-time interactions among entities in the virtual environment, such as collaborative-shopping and shopper-vendor avatar interactions, are implemented over the HLA-RTI.

More HLA-based simulations and applications could be found in the papers of the Simulation Interoperability Workshop[24].

3. Managing the shared spaces through HLA

3.1. Requirements

Singhal and Zyda summarized the common features of networked virtual environments[25], which is also suitable for CVE:

- A shared sense of space: All participants are presented with the illusion of being located in the same place, such as in the same room, building, or terrain.
- A shared sense of presence: When entering the shared place, each participant takes on a virtual persona, called an *avatar*, which includes a graphical representation, body structure model, motion model, physical model, and other characteristics.
- A shared sense of time: Participants should be able to see each other's behavior as it occurs. In other words, the CVE should enable real-time interaction to occur.
- A way to communicate: Though visualization forms the basis for an effective CVE, most CVEs also strive to enable some sort of communication to occur among the participants.
- A way to share: The user's ability to interact with the virtual environment itself and other users.

For a large scale, shared space of CVE, the scalability issues should also be considered. Scalability is the ability of a distributed simulation to maintain time and spatial consistency as the number of entities and accompanying interactions increases[26]:

- Awareness/Visibility Control: The ability of transferring only required data and avoiding transferring non-required data.
- Dynamic Simulation Entity Management: The ability of adding/deleting a simulator into the run-time simulation without negative effects.
- Simulation backup and restore: The ability of recording and retrieving simulation status.

HLA provides a sophisticated infrastructure to satisfy these requirements through the six service groups defined in the HLA Interface Specification:

- Federation Management.
- Declaration Management.

- Object Management.
- Ownership Management.
- Time Management.
- Data Distribution Management.

We are going to illustrate how these services are used to manage the shared space of CVE through the virtual shopping mall example.

HLA is a distributed framework. A federation consists of many federates. Each federate acts a role in the federation. There is a HLA proxy located locally with each federate. The data exchange among federates is completed by HLA proxies and federates do not care about how HLA accomplishes that; refer to Figure 4. From the perspective of the programmer, all HLA services can be accessed through a local call, and all network-involved activities are encapsulated in the HLA-RTI.

3.2. Federation Management (FM)

"Federation Management" refers to the creation, dynamic control, modification, and deletion of a federation execution[15]. The basic federation management activities are shown in Figure 5. Before a federate may join a federation execution, the federation must be created. The first federate is in charge of creating the federation and then other federates may join and resign from it in any sequence.

In the virtual shopping mall application (refer to Figure 6), the shopping mall is a federation, and all customers, dealers, products may be represented as federates. From a technical perspective, customers entering the shopping mall are considered as federates joining the federation. When they leave the shopping mall, the related federates resign from the federation.

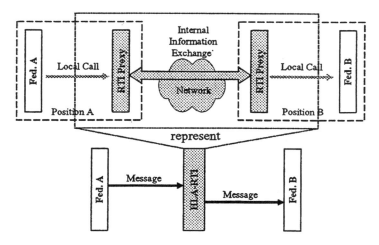

Figure 4. HLA-RTI representation

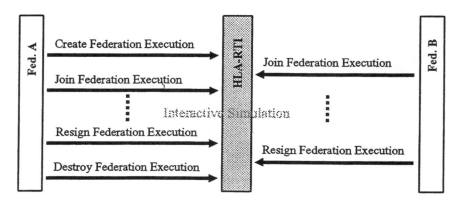

Figure 5. Federation Management – create, join, resign and destroy

3.3. Declaration Management (DM) and Object Management (OM)

Before federates in a federation can see each other and start exchanging information, they must tell the federation what data they can provide to others and what data they need from others. These requirements are handled through Declaration Management and Object Management.

The HLA demands two mechanisms for sharing information between federates: objects and interactions. *Object classes* are comprised of attributes. Object classes describe *types* of things that can *persist*. *Interaction classes* are comprised of parameters. Interaction classes describe *types* of events[17]. The primary difference between objects and interactions is *persistence*. Objects persist, interactions do not. In addition, object attributes can be identified and

Figure 6. Federation Management – virtual shopping mall

updated individually while interaction parameters cannot be identified individually and must be updated based on whole interaction unit. In other words, programmers can update specific object attributes of an object and leave other attributes unaffected but they cannot do the same thing to interactions. When they update interaction, all parameters of the interaction will be affected. Obviously, objects are good at presenting the long live status, while interactions are good for messages.

Declaration Management provides the function to federates to declare their intention to generate and consume information. Object Management deals with the registration, modification, and deletion of object instances and the sending and receiving of interactions[15].

The basic activities of DM and OM services for object classes operations are shown in Figure 7.

Step① indicates that the federate A may subsequently register object instances of object class "C" and provides the information of its attribute "location". Step② indicates that the federate B needs the "location" information of any object instances of object class "C". Similar to the instantiation concept of Object Oriented Programming (OOP), published object classes in HLA also need instantiation before using it and multiple objects can be instantiated from one published object class. In step③, federate A notifies RTI that it creates a unique object instance "x" from the object class "C". When federate A updates the attribute "location" of the object instance "x" in step④, RTI will notify federate B of the new value of the "location" in step⑤.

The basic activities of DM and OM services for interaction classes operations are shown in Figure 8.

The interaction class operation is similar to object class operation. Federates must publish and subscribe their interested interaction class before using it. But an interaction unit does not need registration. The sender federate constructs the interaction, fills all its parameter fields and then sends it out. The receiver federate gets the whole interaction through RTI notification.

In the virtual shopping mall, all avatars declare their position so that they can see each other. Their position data are updated when they are moving, so other

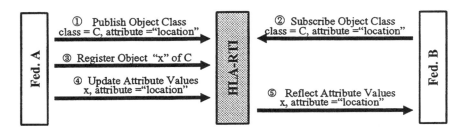

Figure 7. *Declaration Management – object class*

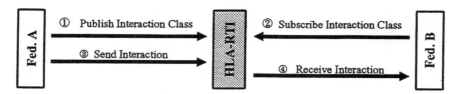

Figure 8. Declaration Management – interaction class

federates can get the new position value and update the customers' browsers respectively; refer to Figure 9.

With the same mechanism, avatars can also declare their other behaviors so that they can interact with each other. For example, the interaction mechanism can be used to transfer short messages among federates. All avatar federates publish and subscribe to the same interaction named "BBS", so any BBS interaction sent out from one avatar federate can be received by all other avatar federates. Then a BBS system is generated.

3.4. Ownership Management (OWM)

Ownership management is used by federates and the RTI to transfer ownership of instance attributes among federates[15]. Only the federate that owns an instance attribute can update the attribute values. The ownership management methods provide a facility for exchanging attribute ownership among federates in a federation

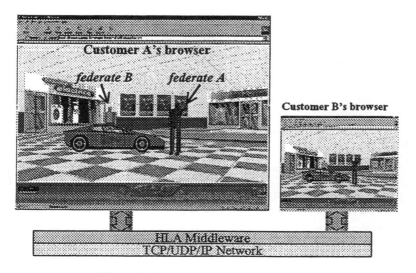

Figure 9. Declaration Management example

execution using a "push" and/or a "pull" model[17]. The push model refers to the fact that a federate can try to give away responsibility for one or more attributes of an object instance; refer to Figure 10. Alternatively, the pull model refers to the fact that a federate can try to acquire responsibility for one or more attributes of an object instance; refer to Figure 11.

A push model of ownership transfer is shown in Figure 10. Federate A owns attribute "location" for object x and publishes updates for the "location". The Federate A requests attribute ownership divestiture of attribute "location" for object x. Then the RTI issues a request for ownership assumption to federate B. Then RTI notifies both A and B that the attribute ownership is exchanged. From this point on, the federate B begins issuing updates for attribute "location" for object x.

A pull model of owership transfer is shown in Figure 11. Federate A owns attribute "location" for object x and publishes updates for the "location" at the beginning. Federate B requests attribute ownership acquisition of attribute "location" for object x. Then the RTI issues a request for attribute ownership Release to the owner of object x, that is federate A. Then the RTI receives the response that federate A will release the ownership of the attribute. Then the RTI notifies that B gets the ownership. From this point, the federate B begins issuing updates for attribute "location" for object x.

To see the utility of ownership exchange in the virtual shopping mall application, refer to Figure 12. Assume that the ownership of the car object belongs to the salesman federate at the beginning. It is very possible that the customer wants to try its options by himself, such as change of color, change of body style, etc. At this situation, the tradesman can transfer the ownership of the car to the customer and then the customer can try it. Of course, if the tradesman does not give up the ownership, the customer cannot get it.

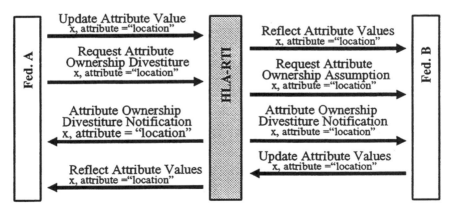

Figure 10. Ownership transfer – push

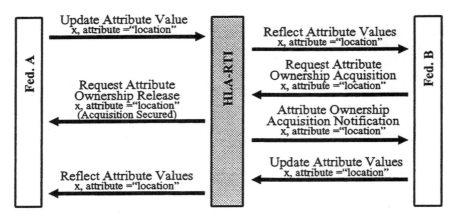

Figure 11. Ownership transfer – pull

3.5. Data Distribution Management (DDM)

The HLA effectively serves as an intelligent switch – matching up producers and consumers of data, based on declared interests and without knowing details about the data format or content being transported. Furthermore, DDM provides a flexible and extensive mechanism for further isolating publication and subscription interests – effectively extending the HLA's switching capabilities.

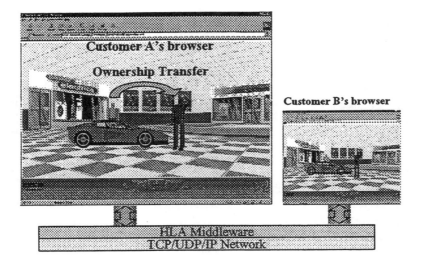

Figure 12. Ownership transfer in the virtual shopping mall

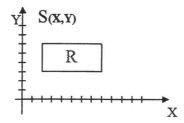

Figure 13. Routing space and region in DDM

In DDM, a federation "routing space" is defined. A routing space is defined as a multidimensional coordinate system and a region is defined as a scope in a routing space. As the example shown in Figure 13, S(X,Y) is a routing space; X, Y are two dimensions of S; R(x:2–8, y:3–6) is a region.

Figure 14 is an example given in the RTI Programmer's Guide[17] about the routing space of DDM which is defined by the three dimensions "longitude", "latitude", and "altitude".

Federates can fine-tune their subscription declarations and data updates in terms of regions within the routing space. Federate A might publish its objects class X within the region R_{Alpha} {longitude: 44°E – 48°E, latitude: 30°N – 37°N, altitude: 0 – 50,000 ft}, and federate B might subscribe to the same objects class X within the region R_{Gamma} {longitude: 40°E – 46°E, latitude: 34°N – 40°N, altitude: 30,000 ft – 50,000 ft}. We can see that the overlap between the two regions $R_{Alpha \cap Gamma}$ {longitude: 44°E – 46°E, latitude: 34°N – 37°N, altitude: 30,000 ft – 50,000 ft} is relatively small. DDM is in charge of judging the overlap and making

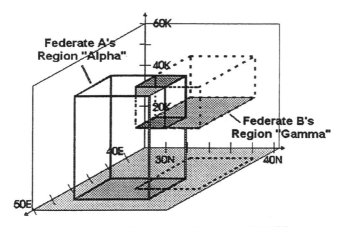

Figure 14. Routing space and region in DDM[14]

sure that only updates fallen into the overlap region will be notified to subscribed federates. For example, the federate A's update to object instance x with value {45°E, 36°N, 40,000ft} will be notified to federate B, while {*43*°E, 36°N, 40,000ft}, {45°E, *39*°N, 40,000ft} and {45°E, 36°N, *10,000*ft} will not be notified to federate B.

Figure 15 shows the declaration object management usage associated with DDM. Federate A offers to publish object class C with attribute "location" with region Alpha. Federate B subscribes to class C and attribute "location" with region Gamma. Federate B can discover the overlap of Alpha and Gamma. When federate A updates the attributes within the overlap, simulator B can get the reflection. When federate A updates the attributes out of the overlap, simulator B is not reflected.

In the virtual shopping mall application, see Figure 16, customers are interested in the products in the store that they visit. They are interested in what the salesman is doing, and even more, they are interested in what the other customers are doing in the same room. But definitely, they don't want HLA to send to their computer all events and status of the whole shopping mall. Obviously, any personal computer cannot handle that. More importantly, customers are not interested in what happens in other stores. Thus the data distribution management service is very important and necessary.

The virtual shopping mall is partitioned into several separate shops, which were defined as different regions in the RTI routing space. Object instances are registered with associate update regions. Then RTI only dispatches update messages to the objects within the same region.

3.6. Time Management (TM)

Time Management is concerned with the mechanisms for controlling the advancement of each federate along the federation time axis[15]. It should be emphasized that the purpose of TM is not to provide a real-world time in federation. Instead, TM provides a synchronization mechanism among federates though time-stamped-ordered (TSO) events and time advances rules.

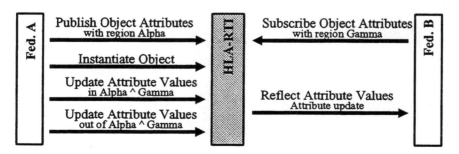

Figure 15. *Data Distribution Management*

Figure 16. DDM in virtual shopping mall

TM is usually used in the distributed simulations that involve heavy duty processing but need real time results or synchronization points. Each federate consumes different simulation time that is far longer than the actual time in the real system. But with the TM, federates can advance time following the actual time in the real system, ignoring the actual simulation time.

Different federates may consume different simulation time. The federates that finish their processing earlier may ask for advancing federation time earlier, but the federation may not grant the time advance and the federates may have to wait. The federation only grants the smallest time request, so earlier finishing federates will wait until the slowest federate finished and requests further time advance; refer to Figure 17.

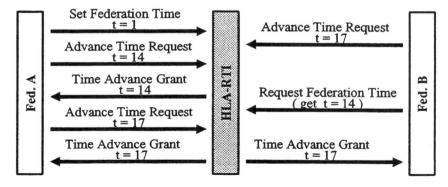

Figure 17. Time Management

In the virtual shopping mall application, all federates are fast enough for real-time response, so no further synchronization mechanism provided by TM is required. But as an example, we can consider a TM-based real-world clock federate in the virtual mall.

The responsibility of the clock federate is to advance the federation time following the real-world time: keeping the current real-world time updated by GPS or other devices; advance the federation time following the real-world time. Any other federates can depend on federation time to update their status or actions respectively, for example, the different light/music/advertisements during different time periods of a day.

3.7. Matches between HLA and requirements of CVE

Reviewing the requirements of CVE, we can see how the HLA matches them.

- The shared sense of space is represented by the federation concept of HLA. The Federation of HLA can provide an illusive space where any simulator can easily join in and resign from.
- The shared sense of presence is supported by the federate concept of HLA with assistance of the Object Management service. Each avatar is implemented as a federate with status and behaviors represented by the attributes of HLA objects.
- The shared sense of time – our virtual shopping mall application verified that HLA-RTI has acceptable performance to support CVE real-time interaction over a LAN environment. Furthermore, the Data Distribution Management service of HLA effectively limits unnecessary events and reduces the response time of a federate. On the other hand, HLA has some limitations for supporting large-scale real-time distributed simulations over the Internet (refer to Section 4.2).
- The way to communicate is supported by the Object Management service of HLA. Objects may be used to represent status updates and interactions may be used to represent messages. Not only simple text-based messages can be exchanged, but also more natural audio communication can be supported[27,28].
- The way to share is supported in HLA through the Objects Management service and the Ownership Management service. The user interacts with the virtual environment itself and with other users by updating object attributes, sending interactions and exchanging ownership of object attributes.
- Awareness/Visibility Control is handled by the Data Distribution Management service. DDM divides the CVE space into smaller regions; the data generated in one region will not be sent to other regions. As a result, awareness and visibility are well controlled.
- Dynamic simulation entity management is supported by the Federation Management service. The Simulation entity is represented as a federate. Any federate can join in and resign from a federation easily at any time by

implementing simple steps. One limitation is that the federate must be predefined in FOM; refer to Section 4.1.

● Simulation backup and restore are handled by the Federation Management (FM) and Time Management (TM) services. FM provides functions for coordinating federation-wide saves and restores. One federate may initiate a federation save or restore and RTI will notify this request to all other federates in the federation. Then all federates save or restore their status. The TM service is used when the optional function parameter – federation time is provided when the first federate initiates the request. The given federation time will be communicated to all federates and all federates will save or restore their status at specific federation time points.

4. HLA limitations for shared space management

As a new DIS standard that was inspired from all old ones, HLA has powerful features to support a very wide range of DIS applications, but there are still two limitations to shared space management.

4.1. Federation Object Model (FOM) issues

HLA demands FOM for each federation. FOM is a profile for concise and rigorous description of the data exchange among federates within a federation, including an enumeration of all object and interaction classes pertinent to the federation. FOM definition is always required to build a federation. All exchangeable data must be declared in FOM. All federates in a federation can only utilize the attributes and parameters defined statically by the FOM.

FOM causes some limitation for CVE. CVE has potential requirements of "openness" which means that a new featured federate should be able to join in after the federation has started. In the virtual shopping mall example, new salesmen and new products may be added to the virtual mall anytime. The new stuffs may have new behaviors or features not defined in the old FOM. To activate these new behaviors and features, the operator has to stop the virtual mall, update the FOM and then start up the virtual mall again. This is inconvenient for a small virtual mall since periodic maintenance becomes necessary; it is also unacceptable for a large scale virtual mall which may be over several time-zones and may have no time for maintenance.

Some experts are looking for alternatives[29]. But we have not found a solution that enables new federates dynamically joining a federation, without predefinition in FOM.

4.2. Quality of Service issues

While there are many soft real-time applications in simulation, there are also numerous applications where a hard real-time feature is an absolute requirement.

Research shows that in a flying simulator not only time delays but also badly timed cues can adversely affect pilot perception, control behavior, and performance[30]. Hard real-time response is not an option or a luxury in this kind of system, but rather an absolute requirement. For example, in a virtual surgery operation that is operated by several surgeons around the world, real-time responses from all the end devices and network are critical.

The real-time requirement is not difficult to match at end nodes by using a real-time operating system. RTI-NG1.3 has a release for VxWorks[31] that can satisfy the real-time requirement of an end simulation node. But for DIS, the network communication delay is also critical for real-time response. For a simulation over a LAN or a campus network, network performance may be controlled through over-configuration. But for a large scale simulation over the Internet, it will depend on the Quality of Service solutions of the Internet. The Internet Engineering Task Force (IETF) is working on QoS over IP. The IETF has identified several solutions through which end-to-end QoS can be realized. HLA has not considered QoS issues. But if HLA wants to support real-time distributed interactive simulations, QoS features must be added to HLA in some way.

5. Conclusion

This paper presented a brief summary of distributed simulation standards, such as SIMNET, ALSP, DIS, SPLINE. It then focused on the High Level Architecture (HLA) standard for Distributed and Collaborative Virtual Environments (CVE) and reviewed in detail its basic management functions for the virtual shared space: federation management, declaration management, object management, ownership management, time management and data distribution management. An example CVE application in e-commerce was viewed under the light of these HLA management functions. Finally, limitations of HLA for shared space management were identified and discussed.

REFERENCES

[1] HLA Homepage. http://hla.dmso.mil
[2] DMSO Homepage. http://www.dmso.mil
[3] OMG Special Interest Group in Simulation.
    http://cgi.omg.org/techprocess/xsigs.html#simsig
[4] HLA clears final hurdle on track to IEEE standardization.
    http://hla.dmso.mil/index.php?page=121
[5] N.D. GEORGANAS, *CRC Report, Industry Canada*, Dec. 1997.
[6] J.I. CLEVELAND, and S.S. HERNDON, *Real-Time Simulation User's Guide*.
    http://www.cs.mcgill.ca/~nader/Red_Book.html
[7] U.S. Army STRICOM, *SIMNET Software Summary*. http://www.stricom.army.mil/
    STRICOM/DRSTRICOM/SOFTWARE/SUMMARIES/simtem.html

[8] ALSP Joint Training Confederation. http://alsp.ie.org/alsp/biblio/mors_96_miller/mors_96.html

[9] E. BERGLUND and H. ERIKSSON, *Distributed Interactive Simulation for Group Distance Exercises on the Web*, Linkoping University, Sweden, 10/26/98. http://www.ida.liu.se

[10] U.S. Army STRICOM, *Advanced Distributed Simulation*. http://www.stricom.army.mil/STRICOM/E-DIR/ES/ADS

[11] ADS & T&E, *Gateway to the Virtual Test Environment*. http://www.jads.abq.com/html/ads/ads.htm

[12] SPLINE homepage. http://www.merl.com/projects/spline/

[13] R.M. WEATHERLY, A.L. WILSON, B.S. CANOVA, E.H. PAGE, A.A. ZABEK. *4. Conclusion, ADS through the ALSP*. http://ms.ie.org/page/papers/hicss-29/camera.html

[14] *Approved Draft 1516.1–2000 IEEE Standard for Modeling and Simulation (M&S) High Level Architecture (HLA) – Framework and Rules*. http://standards.ieee.org/catalog/simint.html

[15] *Approved Draft 1516.1–2000 IEEE Standard for Modeling and Simulation (M&S) High Level Architecture (HLA) – Federate Interface Specification*. http://standards.ieee.org/catalog/simint.html

[16] *Approved Draft 1516.2–2000 IEEE Standard for Modeling and Simulation (M&S) High Level Architecture (HLA) – Object Model Template (OMT) Specification*. http://standards.ieee.org/catalog/simint.html

[17] HLA-RTI Programmer's Guide, p2–1. http://sdc.dmso.mil/

[18] Vendors of HLA Run Time Infrastructure. http://hla.dmso.mil/index.php?page=74

[19] S.L. GRINAWAY, A. LEBIODA, U.S. Army TARDEC, *The Integration of Man-in-the-Loop Distributed Simulation with Motion Simulation Interoperability Workshop*, Spring, 1998. http://www.sisostds.org/doclib/obtain_doc.cfm?record_id=REF_1000349

[20] ROGER D. SMITH, Technical Director, BTG Inc. http://www.modelbenders.com/bio/smithr.html

[21] *Fourth IEEE International Workshop on Distributed Simulation and Real Time Application*. http://www.cs.unt.edu/~boukerch/DS-RT2000

[22] X. SHEN, R. HAGE and N.D. GEORGANAS, "Agent-aided Collaborative Virtual Environments over HLA/RTI", *Proc. IEEE/ACM Third International Workshop on Distributed Interactive Simulation and Real Time Applications (DIS-RT '99)*, Greenbelt MD, Oct. 1999.

[23] J.C. OLIVEIRA, X. SHEN and N.D. GEORGANAS, "Collaborative Virtual Environment for Industrial Training and e-Commerce", *Proc. Workshop on Application of Virtual Reality Technologies for Future Telecommunication Systems, IEEE Globecom' 2000 Conference*, Nov.–Dec. 2000, San Francisco.

[24] Simulation Interoperability Workshop. http://www.sisostds.org/siw/

[25] S. SINGHAL and M. ZYDA, *Networked Virtual Environments – Design and Implementation*. http://www.npsnet.org/~zyda/NVEBook/Book.html

[26] DoD Modeling and Simulation Executive Agent for Terrain, *Modeling and Simulation Terrain Execution Plan, ANNEX 4 DEFINITIONS AND ACRONYMS*, March 21, 1996. http://www.tmpo.nima.mil/new_tep/tep_anx4.html

[27] H. ZHAO and N.D. GEORGANAS, An Approach for Stream Retrieval over HLA-RTI in Distributed Virtual Environments, *Proc. 4th IEEE DS-RT 2000 (Fourth IEEE International Workshop on Distributed Simulation and Real Time Applications)*, San Francisco, Aug. 2000.

[28] H. ZHAO and N.D. GEORGANAS, An Approach for Stream Transmission over HLA-RTI in Distributed Virtual Environments, *Proc. IEEE/ACM Third International Workshop on Distributed Interactive Simulation and Real Time Applications (DIS-RT '99)*, Greenbelt MD, Oct. 1999.

[29] L. GRANOWETTER, MÄK Technologies, Inc. *Solving the FOM-Independence Problem.* http://www.mak.com/tech/fomagile.html

[30] A.D. WHITE, "The Impact of Cue Fidelity on Pilot Behavior and Performance," *I/ITSEC Proceedings*, 1994.

[31] Release Notes of RTI-NG Ver 1.3. http://sdc.dmso.mil

Chapter 2

QoS management for multimedia applications using an efficient queue policy

Seong-Ho Jeong, Henry Owen and John Copeland
School of Electrical and Computer Engineering, Georgia Institute of Technology, USA

Joachim Sokol
Siemens AG, Corporate Technology, Munich, Germany

1. Introduction

With the emergence of real-time applications such as packet voice and packet video, UDP traffic has been increasing over the past few years [THO 97]. In order to support these real-time applications over the Internet, it is necessary to provide a certain amount of bandwidth to the applications within the network so that the Quality of Service (QoS) of the real-time applications will not be seriously affected within the network even during periods of congestion.

The flows of these real-time applications do not typically back off when they encounter congestion, thus they are called unresponsive or aggressive flows. As a result, they aggressively use up more bandwidth than other TCP "friendly" flows that do utilize congestion control. This mix of congestion controlled and congestion uncontrolled traffic could cause an Internet meltdown [BRA 98]. Therefore, while it is important to have router algorithms support UDP flows that require QoS by assigning appropriate bandwidth, it is also necessary to protect responsive TCP flows from unresponsive or aggressive UDP flows and to provide reasonable QoS to all users.

Basically, there are two types of router-based algorithms for achieving a certain QoS: scheduling algorithms and queue management algorithms. Scheduling algorithms can provide sophisticated bandwidth control, but they are often too complex for high-speed implementations and do not scale well to a large number of users. On the other hand, queue management algorithms have had a simple design from the beginning. One of the key examples of queue management algorithms is Random Early Detection (RED) [FLO 93]. A router implementing

RED maintains a single FIFO that is shared by all the flows, and drops an arriving packet randomly during periods of congestion. The drop probability increases as the level of congestion increases. Since RED behaves in anticipation of congestion, it does not suffer from the lock-out and full-queue problems [FLO 93] inherent in the widely deployed Drop Tail (FIFO) mechanism. However, like Drop Tail, RED is not able to penalize unresponsive flows. The resulting percentage of packets dropped from each flow over a period of time is almost the same. As a result, misbehaving flows can continue to use up a large fraction of the link bandwidth and have a serious impact on responsive TCP flows.

To better distinguish unresponsive flows, a few variants of RED such as RED with penalty box [FLO 97] and Flow Random Early Drop (FRED) [LIN 97] have been proposed. However, these algorithms incur extra implementation overhead since they need to collect certain types of state information. RED with penalty box keeps information about unfriendly flows while FRED needs information about active flows. Furthermore, FRED does not consider provisioning of specific bandwidth for real-time UDP-based applications. Recent work in [OTT 99] proposes an algorithm called Stabilized RED (SRED), which stabilizes the occupancy of the FIFO buffer independently of the number of active flows. SRED estimates the number of active flows and finds candidates for misbehaving flows. It does this by maintaining a data structure that serves as a proxy for information about recently seen flows. Although SRED identifies misbehaving flows, it does not propose a simple router mechanism for penalizing misbehaving flows. In addition, it does not consider provisioning of specific bandwidth for real-time UDP traffic.

It has recently been proposed to carry state information in IP packets [STO 99]. The two possibilities to encode state in the packet header are (1) introduce a new IP option or (2) introduce a new header between layer 2 and layer 3, similar to the way labels are transported in Multi-protocol Label Switching. We utilize this technique in this work.

The objective of this paper is to present a simple queue policy algorithm that provides a certain amount of bandwidth to UDP flows that require QoS and also protects TCP friendly flows from unresponsive UDP flows. Our approach also supports drop fairness between TCP flows without maintaining per-flow state. The rest of the paper is organized as follows. Section 2 describes the key features and operations of the proposed queue policy, called threshold-based queue management (TBQM). In Section 3, we provide some simulation results showing that it is possible to meet our goals using the proposed approach. Finally, Section 4 presents a summary of the paper.

2. Algorithm

2.1. Overview

In this section, we describe our proposed queue policy, called threshold-based queue management (TBQM) in further detail. The network model is similar to that used in the Differentiated Services architecture where a network consists of edge routers and core routers as shown in Figure 1.

The edge routers perform packet classification and encode a certain state in packet headers, and the core routers use the state encoded in the packet headers for queue management. The following sections describe the key features of TBQM.

2.1.1. Traffic classification

Since we are focusing on UDP applications that require QoS and TCP applications, TBQM basically supports three different traffic classes: a UDP class that requires QoS (hereafter "QoS-UDP class"), better-than-best-effort-service (BBES) class for TCP traffic, and best-effort-service (BES) class for other traffic. We classify incoming traffic into QoS-UDP class, BBES class, or BES using packet state information inserted in the IP packet header. Note that carrying state in packets is explained in [STO 99]. Initial packet classification is performed at the edge routers which may contact a policy server. Core routers simply check the packet state information to classify incoming packets. Although packets are classified, there is still only one queue of data packets in the core router, shared by all traffic classes.

2.1.2. QoS support for UDP flows and protection of TCP flows

In order to provide high-quality service to the QoS-UDP class, we allocate a certain amount of bandwidth to the QoS-UDP class by assigning a maximum allowable buffer amount for the QoS-UDP class in edge routers and core routers. The necessary buffer size is determined by the agreed upon traffic profile of the specific QoS-UDP traffic which can be determined based upon an admission control procedure or negotiation between a user and a service provider. Suppose that two flows share a finite buffer of size B and are multiplexed onto a link of capacity C using a FIFO scheduler. Flow 1 (e.g., QoS-UDP flow) has peak rate

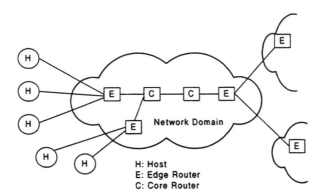

Figure 1. *Network architecture*

specified in its traffic profile while flow 2 is potentially aggressive, and could swamp the first flow if its arrival into the buffer is unregulated. We logically partition the buffer into two portions that correspond to the maximum occupancy levels allowed for flows 1 and 2, respectively, so as to ensure that flow 1 does not lose packets. In this case, flow 1's share of the buffer should be at least as large as its share of the bandwidth, i.e.,

$$B_1 / B \geq \rho_1 / C$$

where B_1 is the flow 1's share of the buffer. It is also shown in [GUE 98] that if flow i requires a guaranteed-service rate ρ_i and is peak-rate conformant, a buffer occupancy threshold of $B\rho_i/C$ is sufficient to guarantee lossless service.

Based on these observations, it is possible to provide a certain amount of bandwidth to QoS-UDP traffic class using a specific buffer occupancy threshold for QoS-UDP traffic class. This also makes it possible to isolate QoS-UDP traffic from TCP traffic by managing each traffic class separately. It should be noted that all packets are still placed in the same buffer using a FIFO approach. The proposed algorithm keeps track and limits how much of the buffer is being used by each of the traffic classes, thus accomplishing the desired goals.

2.1.3. A simple admission control for UDP traffic that requires a QoS

To provide appropriate service to QoS-UDP class traffic, an admission control algorithm is needed to control the admission of QoS-UDP flows based on the availability of resources within the network. We use a simple admission control scheme to decide whether a new QoS-UDP traffic flow is accepted or not. The admission control is based upon the rate information of QoS-UDP flows. The QoS-UDP flows may be sent at constant bit rate (CBR) or variable bit rate (VBR). In the case of CBR QoS-UDP flows, CBR applications simply put the "peak rate" information into the IP header of outgoing packets. In the case of VBR QoS-UDP flows, the edge router estimates the arrival rate of the flows based on the exponential averaging method as in [STO 98]. The calculated rate information is inserted into the IP header of outgoing packets. Note that reference [STO 98] computes flow arrival rates and inserts these rates into packet labels. We utilize this same technique here.

Core routers maintain two variables: the aggregate arrival rate and aggregate accepted rate for QoS-UDP flows. When a special packet, a rate-request packet from an edge router, arrives the core router compares the requested rate with available capacity. If the requested rate is less than or equal to the available capacity, the request will be accepted and the core router forwards the request packet downstream. Otherwise, the request will be rejected, and a reject message will be sent back to the source edge router. Upon receiving the reject message, the edge router does not accept incoming packets that belong to the rejected flow, and it also sends a reject message to the sending host. The sending host would then be made aware that the network is unable to meet its request for bandwidth.

Note that the aggregate accepted rate should be updated because some QoS-UDP flows may have terminated. To do this, the core router calculates an aggregate arrival rate for QoS-UDP flows when a packet arrives. If the aggregate arrival rate is less than the aggregate accepted rate over a certain time interval, the aggregate accepted rate is updated using the largest value of the aggregate arrival rate over the time interval. The aggregate arrival rate A and aggregate accepted rate F are updated using the exponential averaging method [STO 98]. That is,

$$A_{new} = (1 - e^{-T/K})\frac{l}{T} + e^{-T/K}A_{old}$$

$$F_{new} = (1 - e^{-T/K})\frac{l}{T} + e^{-T/K}F_{old}$$

where l is the length of arriving packet, T is the inter-arrival time between the current packet and the previous packet, K is a constant, A_{old} and F_{old} are the values of A and F respectively before the updating.

2.1.4. Metering at edge routers

In order to minimize the impact of QoS-UDP traffic on well-behaved TCP traffic, we use profile meters for the QoS-UDP class traffic. If a QoS-UDP flow is admitted by the admission control procedure described above, we assume that the edge router maintains a traffic profile for the admitted QoS-UDP flow. A traffic profile contains an agreed upon rate between a user and a service provider. The edge router continuously monitors incoming QoS-UDP traffic to check whether or not the incoming traffic violates the traffic profile. Whenever QoS-UDP traffic exceeds the traffic profile, the exceeding traffic is discarded so that TCP traffic will not be affected by the excess traffic. To keep the architecture of core routers simple, profile meters are located in only edge routers. Core routers provide the required bandwidth for the QoS-UDP class traffic by simply maintaining a single buffer occupancy threshold for QoS-UDP class traffic. The core routers only drop QoS-UDP packets as a safety valve in the event that the traffic profile for QoS-UDP at the edge routers does not work properly.

2.1.5. Congestion avoidance and fairness control for TCP flows

We use a simple congestion notification mechanism for TCP flows to avoid congestion. Edge routers insert a special packet, called *choke*, every N packets of each TCP flow. The core router maintains a separate special queue for keeping choke packets for TCP flows. The congestion detection function is performed using an average threshold of the logical TCP queue. The core router maintains statistics about the average queue length using the exponential weighted moving average (EWMA) method. When the average queue length exceeds the threshold, the core router randomly selects a choke packet from the choke queue and sends it back to the source edge router so that the edge router can discard incoming TCP packets based on the received number of choke packets. Since each choke packet contains its own flow ID, it is easy for the edge router to decide which packets to discard. In this manner, packets are discarded only at edge routers, and TCP flows will be dropped fairly based on their rates since the drop rate will be based on the number of received choke packets. This is basically similar to the technique in [VEN 99], but our proposed approach maintains only a single data queue rather than multiple queues.

2.2. Operations

We assume that the core network is provisioned to support a certain amount of QoS-UDP traffic. The admission control procedure determines whether or not new QoS-UDP flows are accepted based on the provisioned capacity and available capacity. If a new QoS-UDP flow is accepted, the edge router keeps a profile for the QoS-UDP flow and continuously monitors the conformity of the flow by comparing the input rate of the flow with the profile. The provisioned capacity is based upon decisions made in a policy-based network environment.

The selection of the amount of queue resources that will be allocated to QoS-UDP class, BBES class, and BES class is not an easy task. Once queue allocation has taken place through traffic engineering, utilization of the network may be measured in part by choke packet activity. If a given core router is generating a large number of choke packets, this is an indication that allocated resources are insufficient. Policy-based network decisions that include the profitability of allocating more resources for a given traffic class at the expense of other traffic classes would need to be a part of this decision.

Edge routers encode class information into the packet state [STO 99] information that is contained in the incoming IP packet. Core routers simply classify incoming packets based on the encoded packet state. Once packets are classified, each core router maintains simple queue statistics for the QoS-UDP class, BBES class, and BES class. The throughput of the traffic classes will be constrained during periods of congestion by limiting the maximum number of packets each traffic class can have queued in the router. Consequently, each class gets only a limited fraction of link bandwidth.

Our queue policy provides isolation between traffic classes by maintaining separate queue occupancy thresholds for each traffic class. These thresholds effectively allocate to each traffic class a portion of the queue's capacity and ensure that this capacity is available to the traffic class independent of the transmission rates of other traffic classes. Packets for a given traffic class are dropped if the queue occupancy of the traffic class exceeds its own queue occupancy threshold.

For the QoS-UDP traffic class, statistics about the number of queued QoS-UDP packets is maintained at core routers. Whenever a QoS-UDP packet arrives, the statistics are updated and compared against a threshold that indicates the allowable maximum number of QoS-UDP packets in the queue. If the updated value exceeds the threshold, the incoming QoS-UDP packet is simply dropped.

Otherwise, the packet is queued. Note that packets are not dropped within the network when the edge router traffic profiles are enforced, and admission control decisions are made correctly.

The congestion avoidance mechanism described in Section 2.1 is applied to TCP packets. That is, if the number of TCP packets in the queue remains less than the maximum threshold of a logical TCP queue, an arriving TCP packet is queued independent of other traffic classes. Otherwise, the core router randomly selects a choke packet from the choke queue and sends it back to the source edge router. The edge router will discard incoming packets that belong to the same flow as the choke packet until the queue length of logical TCP queue is below the threshold. In this way, the early indication for the overutilization of resources in the core network is signaled to the edge router via choke packets. The edge router then adjusts rates so as to prevent too many packets from arriving at a core router. Thus, packet drops occur before the core network resources are congested.

3. Simulations

In order to demonstrate the effectiveness of TBQM, we used simulations to compare the performance of TBQM with simple drop-tail (FIFO) queuing, RED, and FRED. Since the focus of our simulations was to show how well TBQM behaves to support UDP traffic as well as TCP traffic, we only considered the QoS-UDP class and BBES class. Note that our approach can be extended to support more classes by using multiple buffer occupancy thresholds.

The simulated network topology is shown in Figure 2, where there are four TCP sources and five UDP sources, three routers, and a sink. The TCP traffic originates from a collection of ftp sources, and the UDP traffic originates from a set of CBR sources. The edge router maintains traffic profiles for UDP sources after the admission control procedure is complete as described in Section 2.1.3. Each UDP source is allowed to send 1 Mbps UDP traffic. The edge router continuously monitors incoming UDP traffic to check if the input traffic violates the traffic profile.

In our simulations (we used the network simulator NS), we considered a single congested link between two core routers, which is shared by all flows. Specifically, a core router is configured with a 10 Mbps outbound link that is the bottleneck link in the network. When there is no congestion on the link, TCP users are capable of utilizing about 62% of the 10 Mbps link, in aggregate.

It is assumed that the minimum guaranteed rate for each UDP user is 1 Mbps. Since there are five UDP users, it is necessary to reserve 5 Mbps, in aggregate, in the network. The total buffer size of FIFO queue is 30 KB. Based on the discussion in Section 2.1.2, the buffer occupancy threshold for QoS-UDP traffic is calculated as 15 KB. Accordingly, the available buffer size for other traffic is 15 KB. RED and FRED as well as a FIFO queue management scheme were simulated for comparative purposes. The values of min_{th}, max_{th}, w_q, max_p for RED and FRED are 15 KB, 30 KB, 0.002, and 0.02, respectively.

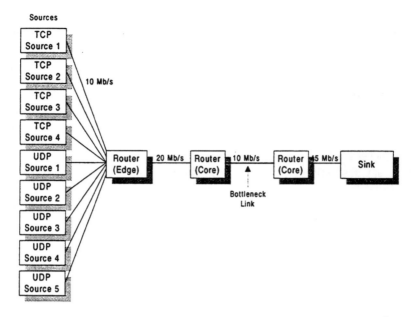

Figure 2. *Network topology for simulations*

Figure 3 illustrates the TCP throughput measurements on the outbound link of the congested router. In each experiment, a different queue management scheme is employed in all routers including edge and core routers. We ran infinite TCP sources during simulations. At time 30 s, all UDP sources begin to generate unresponsive QoS-UDP traffic at the rate of 1 Mbps each. Thus, the total aggregate input rate of UDP sources is 5 Mbps from time 30 s to 50 s. At time 50 s, UDP source 5 additionally generates 3 Mbps unresponsive QoS-UDP traffic until time 70 s. As a result, the total aggregate input rate of UDP sources is 8 Mbps from time 50 s to time 70 s. All UDP sources stop generating traffic at time 70 s.

As shown in Figure 3, the aggregate throughput of TCP traffic is approximately 6.2 Mbps for each of the four methods until time 30 s. After five UDP sources are turned on at time 30 s, the aggregate throughput of TCP traffic in all scenarios is reduced to approximately 5 Mbps. Congestion begins at time 30 s since the total input rate (5 + 6.2 = 11.2 Mbps) exceeds the link capacity of 10 Mbps. In our simulations, choke packets were not used to reduce congestion so that we could show what happens when there is congestion. Note that RED shows worse TCP performance during this time interval since early discarding of TCP packets reduces the TCP input rate, and RED does not protect TCP flows from unresponsive QoS-UDP flows.

At time 50 s, congestion is more severe since the total aggregate input rate is increased to 14.2 Mbps. As shown in Figures 3(b), (c), FIFO and RED show the worst TCP performance since unresponsive QoS-UDP traffic is not punished and consumes most of the link capacity. The results in Figure 3(d) show that FRED protects TCP flows from QoS-UDP flows, however FRED does not guarantee the minimum rate (5 Mbps) for QoS-UDP traffic. On the other hand, as shown in Figure 3(a), TBQM shows better performance than other schemes in terms of TCP flow protection and minimum rate guarantee for QoS-UDP flows. That is, from

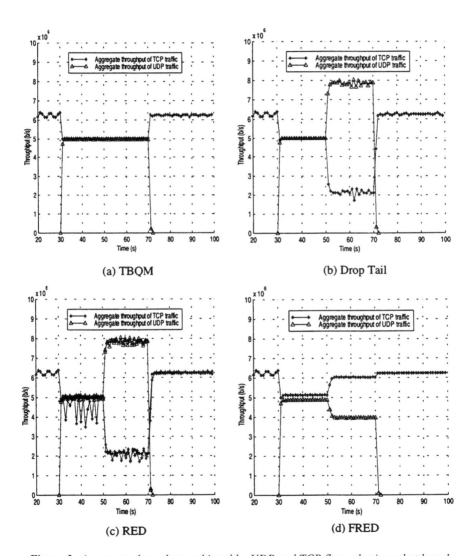

(a) TBQM

(b) Drop Tail

(c) RED

(d) FRED

Figure 3. Aggregate throughput achieved by UDP and TCP flows sharing a bottleneck link of capacity 10 Mbps in the network

time 50 s to 70 s, TCP traffic is not affected by excess QoS-UDP traffic, and TCP gets approximately 5 Mbps continuously. Furthermore, QoS-UDP traffic gets the desired minimum rate (5 Mbps) during the 50 to 70 second simulation time period.

4. Conclusions

The volume of multimedia traffic in the Internet is dramatically increasing. The current best-effort forwarding model of the Internet is frequently insufficient for supporting multimedia traffic requirements. For example, the current Internet does not efficiently support UDP-based real-time applications such as voice over IP and video conferencing.

To satisfy the performance requirements of these ever more common applications, it is necessary to provide a certain amount of bandwidth within the network so that the performance of the applications will not be seriously affected during network congestion. Since the flow rates of some of these applications do not back off during periods of congestion, it is also necessary to protect responsive TCP flows from unresponsive UDP flows. To achieve these goals, we proposed a queue policy, called threshold-based queue management (TBQM). TBQM efficiently isolates UDP flows from TCP flows to protect TCP flows. TBQM does this by using logical buffer partitioning so as to support bandwidth requirements of UDP applications by reserving buffer space for UDP traffic.

To support our queue policy, we also proposed a simple admission control procedure for UDP traffic and a drop fairness control scheme for TCP traffic during periods of congestion. To demonstrate the effectiveness of TBQM, we compared the proposed approach with Drop-Tail (FIFO), RED, and FRED using simulations. TBQM showed better performance than other schemes in terms of TCP flow protection and minimum rate guarantee for UDP flows.

REFERENCES

[BRA 98] BRADEN, B., CLARK, D., CROWCROFT, J., DAVIE, B., DEERING, S., ESTRIN, D., FLOYD, S., JACOBSON, V., MINSHALL, G., PARTRIDGE, C., PETERSON, L., RAMAKRISHNAN, K., SHENKER, S., WROCLAWSKI, J., ZHANG, L., "Recommendations on Queue Management and Congestion Avoidance in the Internet", *IETF RFC (Informational) 2309*, April 1998.

[FLO 93] FLOYD, S. and JACOBSON, V., "Random Early Detection Gateways for Congestion Avoidance", *IEEE/ACM Transaction on Networking*, vol. 1, no. 4, p. 397–413, August 1993.

[FLO 97] FLOYD, S., and FALL, K., "Router Mechanisms to Support End-to-End Congestion Control", *LBL Technical Report*, February 1997.

[GUE 98] GUERIN, R., KAMAT, S., PERIS, V., and RAJAN, R., "Scalable QoS Provision Through Buffer Management", *Proceedings of ACM SIGCOMM'98*, September 1998.

[LIN 97] LIN, D., and MORRIS, R., "Dynamics of Random Early Detection", *Proceedings of ACM SIGCOMM'97*, p. 127–137, October 1997.

[OTT 99] OTT, T., LAKSHMAN, T. and WONG, L., "SRED: Stabilized RED", *Proceedings of INFOCOM'99*, p. 1346–1355, March 1999.

[STO 98] STOICA, I., SHENKER, S., ZHANG, H., "Core-Stateless Fair Queueing: Achieving Approximately Fair Bandwidth Allocations in High Speed Networks," *Proceedings of ACM SIGCOMM'98*, September 1998.

[STO 99] STOICA, I. and ZHANG, H., "Providing Guaranteed Services Without Per Flow Management," *Proceedings of ACM SIGCOMM'98*, September 1999.

[THO 97] THOMPSON, K., MILLER, G., WILDER, R., "Wide-Area Internet Traffic Patterns and Characteristics," *IEEE Network*, vol. 11, no. 6, November/December 1997.

[VEN 99] VENKITARAMAN, N., SIVAKUMAR, R., KIM, T., LU, S., BHARGHAVAN, V., "The Corelite QoS Architecture: Providing a Flexible Service Model with a Stateless Core," *Working Draft of UIUC*, February 1999.

Chapter 3

Structuring devolved responsibilities in network and systems management[1]

E. Lupu, N. Dulay, N. Damianou and M. Sloman
Department of Computing, Imperial College, London, UK

1. Introduction

The size and complexity of distributed systems and networks has grown exponentially. Modern networks must become more adaptable to cater for the wide range of user devices ranging from powerful multi-media workstations to web-enabled cellular telephones. There is a need to support fast service creation and resource management through a combination of network-aware applications and application-aware networks. Services will increasingly be provided by multiple layers of 'value-added' service providers. Management systems must also be flexible and adaptable and provide access to management facilities by clients who may be third party service providers or end-users rather than just the provider's administrators. The management framework must make explicit the devolution and distribution of responsibility for both human and automated agents necessary to cater for large-scale, multi-organisational systems.

The management functions provided to third party providers and clients could intentionally or inadvertently be misused. Users' access must therefore be controlled in terms of the exact operations they are permitted to perform on managed objects, in order to have a clear and analysable specification of *authorisation policy*. Furthermore, there is a need to clearly specify the responsibilities assigned to managers, i.e., the actions that human and automated agents are *obliged* to perform in response to events or at specified times. We define *policy* as a *rule governing the choices in behaviour of the system*. Our policies are interpreted so they can be dynamically modified to change management behaviour and strategy. Policies specify which actions have to be performed and when, but not what the actions consist of, i.e., their implementation. They are specified in

---

[1] This paper is based on E. Lupu, N. Dulay, N. Damianou and M. Sloman, Ponder: Realising Enterprise Viewpoint Concepts, *Proc. 4th Int. Enterprise Distributed Object Computing (EDOC2000)*, Makuhari, Japan, 25–28 Sept. 2000, IEEE Press, pp 66–75.

accordance with business goals and then refined until they can be implemented by manager agents distributed across the network. Note that agents interpreting policies can be elastic processes able to expand their capabilities by dynamically loading programs from various sources.

This paper investigates the structuring and control of delegated responsibilities using policies and role-theory concepts [1]. The operation of a GSM network is used as a scenario to illustrate the concepts presented in this paper. Many of the example roles relate to human managers as these are often more intuitive, but the framework caters for both human and automated management agents. After a brief outline of the scenario (Section 2) we recap the essential aspects of management policies and domains (Section 3). In Section 4 we introduce role concepts, specifications and relationships, while Section 5 explains how to define organisational structures in terms of instances of roles and relationships. Section 6 gives a brief overview of constraint specification. We discuss the related work in Section 7 followed by conclusions.

2. Scenario overview

Our scenario is the operation of an imaginary GSM network operator, RCom. We limit ourselves to the management of Base Station sub-systems (BSS) and subscriber data and assume the network is divided into regions with branches.

A number of roles are used for the management of the network at the branch level. The Help-Desk (HD) role is responsible for interacting and supporting the customers of the company. The Telephone Service Engineer (TSE) role encompasses the investigation of faults occurring on the radio interface between mobile stations and base transceiver stations and the determination of whether a base network operator should be alerted to deal with the fault. Base Network Operators for Switches (BNoS) are responsible for managing the Mobile Switching service Centre (MSC) and the Visitors Location Register (VLR) situated in a branch and cooperate with the BNoRs for radio-related functions. On the other hand, Base Network Operators for Radio (BNoR) are responsible for Base Transceiver Systems (BTS) and cooperate with the TSEs. Finally, Network Element Administrators (NEA) perform all on-site management tasks requested by BNoS and BNoR. They may initiate tests and configuration procedures defined in the Operations Management Centre Workstations (OMC). In this paper we focus on the TSE, BNoS and BNoR roles.

The management components for our network are shown in Figure 1. We assume an object interface to network components and abstract events instead of TMN concepts for management and SS7 for network signalling [5].

3. Management framework

Our role framework is based on the use of management policies and the ability to group objects into domains for applying a common policy. This section recapitulates the main aspects relating to domains and policy specification.

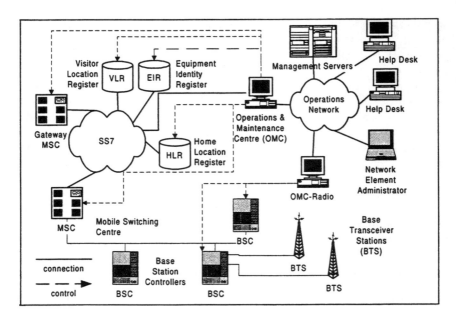

Figure 1. *Mobile cellular network overview*

3.1. Domains

In large-scale systems there is a need to group objects in order to apply a common policy, to partition management responsibility and authority, to reflect geographical or organisational boundaries or for the convenience of human managers. For example, RCom's national network could be divided into 4 or 5 regions, corresponding to different local carriers that maintain their own Home Location Registers (HLRs). Each region could be further sub-divided into a number of branches, each corresponding to the coverage area of a MSC (800,000 inhabitants according to [14]).

Domains [18,19] are used to group references to objects and sub-domains in a structure similar to that of a file system directory. However, domains may hold references to any type of object including people. Furthermore, objects can be members of multiple domains i.e., domains can overlap. Domains have been implemented as directories in an extended LDAP Service.

The domain structure for the Wales region is shown in Figure 2 with one branch (**Cardiff**) expanded into sub-domains to group roles, policy objects, relationships, the operations network (**opn**) and the telecommunications network (**tn**). The telecommunications domain for a branch is further elaborated into sub-domains containing the MSC, VLR and the Base Station Controllers (BSC). In addition to the branch sub-domains, the root domain for a region also has a region-wide telecommunications network domain that includes the tn sub-domains of all

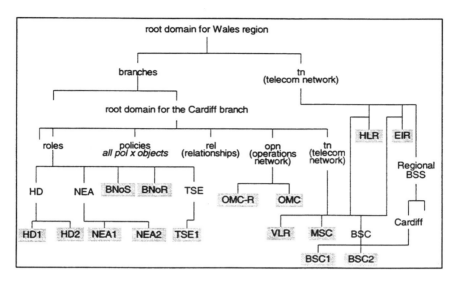

Figure 2. Regional domain structure (shaded boxes represent leaf objects)

branches as well as domains for the Home Location Registers (HLR) and Equipment Identity Registers (EIR). The region-wide telecommunications network domain includes a Base Station Subsystem (BSS) sub-domain which holds each branch's BSC sub-domains (branch tn domains are thus overlapping with the regional BSS domain). This allows regional policies, implemented by managers in a regional operations centre to be defined for the uniform configuration, or logging strategy of all BSCs in the region.

Path names can be used to identify domains and objects in the domain structure e.g., /wales/cardiff/tn/VLR refers to the VLR of the cardiff branch in region wales of the national network (root domain – '/'). Policies are specified in terms of domains, so objects can be added and removed from domains without having to change the policy specifications. Policies applying to a domain normally propagate to members of sub-domains so a policy applying to /wales/cardiff/tn will also apply to all BSCs managed within the branch.

3.2. Management policies

Policies [18] are rules governing the choices in behaviour of management agents in a system. They define an agent's responsibilities, in terms of the actions it must perform, and determine its rights to access system resources. A precise notation that can be analysed and checked for inconsistencies [9] is needed to specify the rights and duties of both human and automated agents in the system. This paper focuses on the Ponder language [3] for specification of policies and their grouping into roles. Issues relating to policy implementation and conflict analysis have been presented in [9, 11].

Authorisation policies define what activities subjects can perform on target objects and are essentially access control policies to protect resources from unauthorised access. Constraints can be specified to limit the applicability of both authorisation and obligation policies based on time or values of the attributes of the objects to which the policy refers.

```
inst auth+ ×1 {
    subject  cardiff/roles/HD;
    target   cardiff/tn/HLR;
    action   addSubscriber(), removeSubscriber(), updateRecord(),
             lockSubscriberInHLR();
}
```

Policy ×1 states that Help-desk staff in the cardiff branch are authorised to add, remove or update subscriber information in the HLR (policy target). They are also authorised to lock subscribers to temporarily suspend their usage of the account.

```
inst auth– ×2 {
    subject  staff = cardiff/roles/HD;
    target   cardiff/tn/EIR;
    action   blacklistEquipment();
    when     staff.status = "trainee";
}
```

Policy ×2 states that Trainee help-desk staff are forbidden to blacklist equipment in the EIR.

Obligation policies define what activities managers or agents must perform on target objects. Obligation policies are triggered by events.

```
inst oblig ×3 {
    subject  op = cardiff/roles/HD;
    target   cardiff/tn/HLR;
    on       newServiceSubscription(IMSI, serviceID);
    do       updateRecord (IMSI, serviceID) → op.log("update", IMSI,
             serviceID);
    when     time.between ("0800", "1900");
}
```

The positive obligation x3 is triggered by an external event signalling a new service subscription from a customer denoted by its subscriber identity (IMSI). The help-desk operator, or an automated agent on its behalf, must update the customer's record in the HLR and log the update (e.g. to an internal log file). While the updateRecord operation is defined on the HLR object, the log operation may be an internal operation in the agent. These operations may only be performed during office hours as specified by the constraint.

```
inst refrain ×4 {
    subject   eng = cardiff/roles/TSE;
    target    cardiff/tn/VLR;
    action    traceForeignSubscriberInVLR ();
    when      eng.state = "standby";
}
```

The refrain policy x4 states that telephone service engineers must not initiate traces in the VLR when they are in the standby state, i.e. a constraint on the internal state of subjects.

More details on the policy notation can be found in [2, 3]. Policies applying to an enterprise system may be abstract and need to be refined into concrete realisations in order to be implemented. The progressive refinement of an abstract policy or goal leads to a refinement hierarchy where lower levels realise the more abstract representation above. High-level abstract policies can thus be refined into implementable policies. To record this hierarchy, policies automatically contain references to their parent and children policies.

Policy types are needed to instantiate multiple policies for different managers in various branches of RCom's network, with similar responsibilities for different managed objects. Policy types can use parameters for defining subjects or targets or both, which permits reuse of policy specifications. For example, telephone service engineers (TSE) must initiate traces on visiting Mobile Stations (MS) grey-listed in their home network for suspected illegal activities. Different instances can be created from this type by specifying the subjects and targets. For example, in the cardiff branch, the subject would be instantiated to cardiff/roles/TSE and the target to cardiff/tn/VLR.

```
type oblig init_trace (subject tse; target ms) {
    on greyListUse(IMEI, IMSI);
    do traceForeignSubscriberInVLR(IMEI, IMSI);
}
```

```
inst oblig cardiff_trace = init_trace (cardiff/roles/TSE, cardiff/tn/VLR);
```

4. Roles

4.1. Role concepts

Specifying organisational policies in terms of role-positions (e.g. branch network operator or regional network manager) rather than persons, permits the assignment of a new person to the position without re-specifying the policies. A role is thus the position and the set of authorisation and obligation policies defining the rights and duties for that position (Figure 3). A role can also be used to group the authorisation and obligation policies that determine the functions performed by automated agents e.g., a configuration or monitoring agent. The

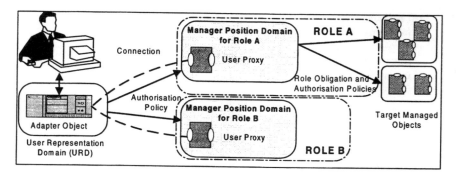

Figure 3. *Role assignment*

implementation of role-specific functions can be dynamically loaded from a repository thus permitting reuse of the underlying agent implementation.

A *User Representation Domain (URD)* is a persistent representation of a user within the system. When a user logs-in, an adapter object is created within the URD to act as an interface between the human and the system. Policies applying directly to the user, e.g., authorisations to access private files, have the URD as subject and thus propagate to this adapter object.

Role policies have a common subject domain called the *Manager Position Domain (MPD)*. A user is assigned to the role by authorising the adapter object executing on behalf of the user in the URD, to connect to a proxy object in the MPD which acts as the user's representative in that role (Figure 3). This ensures that the user can be assigned to several roles while preserving the separation of context between the user's actions in each of them. The adapter object can be thought of as being similar to an X server providing separate windows to the proxy objects in each role. A proxy managing a window may customise the menus provided to the user according to the authorisation policies specified within the role.

4.2. Role specification

The rights and duties for telephone service engineers in the operation of a GSM network can be written as follows:

```
type role engineer (vlr, eir) {
    /* On a handset failure the engineer must update the record in the
    Equipment Identity Register (eir) */
    inst oblig update {
        target  eir;
        on      hs_failure (x);
        do      eir.update_record (x); }
```

```
/* On use of grey-listed MS initiate a trace in the visitors location register
(vlr) */
inst oblig trace {
    target  vlr;
    on      grey_list_use (x);
    do      vlr.trace_foreign (x); }

/* other policies */
}
```

A role type may include multiple policy and constraint instances as well as further type definitions and can be extended by specialisation. For example, a common role type definition can be specified for base network operators grouping their common rights and duties:

```
type role op (set t) {

    /* Restart failed equipment in target domain */
    inst oblig restart {
        target  f = t ^ {id};
        on      failure (id);
        do      f.restart () → f.run_self_test (); }

    /* other policies */
}
```

This specification can be extended to define the role types for base network operators for switches (switch_op) and for radio equipment (radio_op) with the specific rights and duties of each. Specialisation of a role type is achieved by the keyword extends followed by the name of the role type which is extended, and the parameter values for that role.

```
type role switch_op (set <bsc_type> b, <msc_type> m) extends op (b) {

    /* On a circuit failure the circuit must be blocked and reset. */
    inst oblig reset {
        on      A_failure (x, bsc);
        target  f = b ^ {bsc};
        do      f.reset (x);}

    inst oblig block {
        target  m;
        on      A_failure (x, bsc);
        do      m.block_circuit (x, bsc);}

    inst auth+bsc_auth {
        target  b;
        action  reset, disable;}

    inst auth+msc_auth {
        target  m;
        action  block_circuit, unblock_circuit; }
}
```

type role *radio_op* (**set** *<bsc_type>* b) **extends** *op* (b) {

   inst oblig reset {...}

   inst auth+ auth_station_controller {
      target b;
      action force_handover, bar_cell, shutdown; }

   /* On cell-overload, force a hand-over of all connected mobiles */
   inst oblig clear_cell {
      on   cell_overload (bts, bsc);
      target f = b ^ {bsc};
      do   f.force_handover (bts); }

   inst oblig fail_reconfigure {
      on   failure (cir, bts, bsc);
      target f = b ^ {bsc};
      do   f.disable (bts, cir) → f.enable (bts, "backup"));}

   /* On 3 occurrences of a failure event report, increase BTS transmission */
   inst oblig increase {
      on   3*radio_link_fail (bts, bsc);
      target f = b ^ {bsc};
      do   f.increase_power(1);}
}

4.3. Role relationships

Individuals assigned to roles do not work in isolation but collaborate and interact with each other in order to perform their tasks. *Relationships* define the rights and duties of the related roles towards each other, e.g., the right to assign a task to a role, as well as the interaction protocols for collaboration between the managers assigned to those roles. For example, requesting an on-site intervention by a Network Element Administrator (NEA) must be done according to a protocol which ensures that all assignments are logged and leaves no ambiguity as to whether the intervention has been completed, is in progress or cannot be performed. Thus, a relationship is composed of a number of participants (roles), a set of interaction protocols, a set of policies applying to the participants, and a set of constraints applying to the policies. Details on the role interaction protocol can be found in [7].

A relationship is defined in Ponder by specifying the roles that participate in the relationship and the policies associated with these roles. Policies that define the rights and duties of the roles towards each other have the relevant roles as subjects and targets. For example, in the case of our GSM operations scenario, base network operators collaborate with network element administrators and are authorised to assign them tasks. This assign_rel relationship type is specified below, and further specialised for the cases of a BNoS (switch_repair) and a BNoR (radio_repair) using the same inheritance mechanism as for roles.

```
type rel assign_rel (administrator admin, op operator) {
    inst auth+assign_pol {
    subject    operator;
    target     admin;
    action     assign;}
}
type rel switch_repair (administrator admin, switch_op operator)
    extends assign_rel (admin, operator) {
        ...
}
type rel radio_repair (administrator admin, radio_op operator)
    extends assign_rel (admin, operator) {
        ...
}
```

This assign_rel relationship comprises the assign_pol authorisation policy instance which gives base network operators the right to assign tasks to network element administrators. The subject and target of this policy are the participant roles of the relationship as defined by the role statement. Note that these roles are typed. Hence, an instance of this relationship can be created only between roles of types administrator and op respectively. This relationship type is further specialised and these extended versions require participant roles to be of types switch_op and radio_op which are sub-types of op. This demonstrates how strong typing principles from classical object-oriented programming have also been applied in Ponder in order to maximize the compile time verification and ensure rigorous specifications.

5. Organisational structure

An organisational structure is formed by a number of roles interconnected by relationships reflecting authority, supervision, collaboration, delegation, etc. [12]. For the management of the cardiff branch we will consider that only one BNoS as well as one BNoR, one TSE, two NEA and two HD are needed with relationships as shown in Figure 4. The number of role instances and the relationships between them are decided according to the particular requirements of the branch and can be changed dynamically according to changes in the infrastructure or in the number of subscribers.

The branch management structure, shown in Figure 4, is itself sub-divided in two management structures – one responsible for the customer care and the other responsible for the management of network elements. The customer care structure groups the help-desk roles and the telephone service engineer role that is responsible for investigating failures. Network element administrators and base network operators are grouped in the network element management structure. This configuration can be specified in Ponder as shown below.

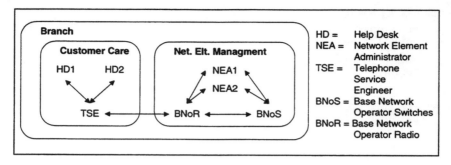

Figure 4. *Role instances and relationships between them*

type mstruct *branch* (**domain** br, nw) {
    import/type/*custcare*;/type/*netelements*;

    inst mstruct cc = *custcare* (br, nw);
                ne = *netelements* (br, nw);

    type rel *radiofail* (**role** eng, radio_op) {...}

    inst rel f = *radiofail* (cc.tse, ne.bnor);
}
domain c = .../wales/branches/cardiff/;
inst mstruct cardiff = *branch* (c, c/tn/);

The branch type, declared with the keyword mstruct takes two parameters, the domain br in which it resides and the domain nw in which the various network elements reside. Like any other composite type in Ponder, an mstruct can contain any number of type declarations, and any number of instances, declared as two blocks with the keywords type and inst respectively. Ponder does not impose limits on the number of these blocks or their ordering. In this example, type specifications for the custcare and netelements mstructs have been imported from a type repository and one instance of each has been created within the branch. The branch mstruct type also contains the declaration of a relationship type and its instantiation. The relationship instance f specifies the policies governing the behaviour of the TSEs and of the BNoRs towards each other. An mstruct may contain roles, policies, groups of policies, relationships and other mstructs. Finally, an instance of the branch is created for the Cardiff area. Similarly we specify below the mstruct types for the custcare and netelements structures from Figure 4.

type mstruct *custcare* (**domain** br, nw) }
    import   /type/*helpdesk*;/type/*engineer*;/type/*complaint*;

    domain   rs = br/roles/tse; h = br/roles/hd/; s = br/rel/;

    inst role   h/hd1 = *helpdesk* (...);
                     h/hd2 = *helpdesk* (...);
                     rs/tse1 = *engineer* (nw/vir, nw/eir);

```
    inst rel    s/hd1_tse = complaint (hd1, tse);
                s/hd2_tse = complaint (hd2, tse);
}
type mstruct netelements (domain br, nw) }
    import      administrator, switch_op; radio_op; repair; base;

    domain      r = br/roles/; a = r/nea/; s = br/rel/;

    inst role   a/nea1 =   administrator (...);
                a/nea2 =   administrator (...);
                r/bnos =   switch_op (nw/bsc/, nw/msc);
                r/bnor =   radio_op (nw/bsc/);

    inst rel    s/bnos_nea1 = switch_repair (bnos, nea1);
                s/bnos_nea2 = switch_repair (bnos, nea2);
                s/bnor_nea1 = radio_repair (bnor, nea1);
                s/bnor_nea2 = radio_repair (bnor, nea2);
                s/bnos_bnor = base (bnos, bnor);
}
```

6. Constraints

Ponder uses a subset of the Object Constraint Language (OCL) [15] for specifying constraints which limit the applicability of a policy, for example to a particular time interval or according to the state of the system. These are specified as part of the individual obligation, authorisation and refrain policies. Semantic constraints which restrict the set of permissible policies (also referred to as Meta-Policies) can be specified as part of any composite policy including roles, relationships and management structures [9]. These are needed in order to determine application-specific conflicts such as separation of duties or conflicts for resources. For example, a conflict of separation of duties can be specified in Ponder as follows:

```
type meta dutyconflict (act1, act2, targ_type) raises dcevent (z) {
    [z] = self.policies → select (pa, pb |
            pa.subject → intersection(pb.subject) → notEmpty       and
            pa.action → exists(act | act.name = act1)              and
            pb.action → exists(act | act.name = act2)              and
            pb.target → intersection(pa.target) → oclIsKindOf (targ_type)
        );
    z → notEmpty
}
```

This constraint can be instantiated in order to specify that the same users are not permitted to authorise a payment and initiate it:

`inst meta` dc = *dutyconflict* ("authorise", "initiate", "payment");

or that help-desk staff are not allowed to both update payment details and service subscriptions on customer records:

`inst meta` ss = *dutyconflict* ("serviceUpdate", "paymentUpdate", "customerRecords");

7. Related work

A number of related studies on policies for network and distributed systems management exist [10]. There is considerable interest in the Internet community in using policies for bandwidth management where typically policy enforcement agents such as routers delegate the processing of policy decisions to policy servers that hold policies as objects in directories [10]. On receiving a policy decision from a policy server, the policy client then implements the decision, for example by forwarding or dropping a packet or allocating resources for subsequent packets. The IETF policy framework [13] has no clear separation of authorisation policies and obligation policies; rather IETF policies are defined as actions that are performed when some condition becomes true. IETF policies can be modelled directly in our system as event-triggered obligation policies with constraints used to further limit the policy, although a better approach would be to attempt to identify authorisation policies and model them explicitly. The IETF uses two separate concepts for grouping: directories for holding policy objects, and dynamic groups for specifying subjects and targets, the latter being specified either by enumeration or by a predicate. In Ponder, we use domains for grouping policies, and for grouping subjects and targets, and if needed use policy constraints to further limit the scope of a domain used in a policy. The explicit separation of subjects and targets in policies allows us to partition and distribute policy information only to where it is needed while the use of domains for grouping policies allows us to use Ponder for self-enforcement and self-management, i.e. to define and enforce policies about policy objects.

The Policy Description Language (PDL) [20] is an event-based language originating at the network computing research department of Bell-Labs. Policies in PDL are similar to Ponder obligation policies. They use the event-condition-action rule paradigm of active databases to define a policy as a function that maps a series of events into a set of actions. The language has clearly defined semantics, and architecture for enforcing PDL policies has been specified. The language can be described as a real-time specialised production rule system to define policies. Events can be composite events similar to those of Ponder obligation policies. PDL does not support access control policies, nor does it support the composition of policy rules into roles, or other grouping structures.

The Enterprise Viewpoint of the Open Distributed Processing Reference Model (ODP-EV) considers *communities* as the primary building blocks of the organisational structure [4]. In essence, a community is an n-party relationship formed to meet an objective where the behaviour of the participants, i.e., their role within the community, is governed by a set of policies. The organisational structure is then described as a hierarchy of communities. However, there is a need to specify roles which participate in different communities, i.e., relationships. Whilst in our model the separation between roles and relationships permits roles to participate in multiple relationships the ODP-EV needs to consider communities which overlap in their role set. This breaks the encapsulation communities provide. Also, the ODP-EV does not provide a policy notation or a model of interactions in relationships. The relationship between our policy concepts and those of the ODP viewpoint is discussed in [6].

Sandhu [16] uses role inheritance to support the re-use of permissions in order to model organisational structures where more "senior" roles inherit the access rights of "junior" roles. Although this model may appear attractive, it violates organisational control principles where access rights should not be inherited by senior roles from junior roles [12] and typically leads to a large number of exceptions in the definitions of derived roles in real-world situations. A more detailed comparison between our work and RBAC models is presented in [8], and a study on the different types of role hierarchies in RBAC is presented in [12].

8. Conclusions

Getting the right domain structure which reflects the organisational requirement for object groupings and what objects need to be shared is an important starting point in managing enterprise-wide distributed systems. Identifying roles in terms of rights and duties associated with positions is the next stage. Providing a representation for roles within the management system permits the distribution of responsibilities reflecting organisational structure as well as analysis for auditing and review purposes. Role classes permit the reuse of policy specifications and reflect the categorisation of classes of employees found in many organisations. Managers assigned to roles do not work in isolation but collaborate in a complex network of various types of relationships. Our role framework supports specification of the rights and duties imposed upon the related parties and the interaction protocols needed for collaboration.

The framework presented in this paper has also showed how representing the requirements for role collaboration in terms of associations between role and relationship classes can be used to define configurations or patterns that can be repeated in multiple branches or departments in an organisation.

We are currently developing a policy specification toolkit to aid administrators in managing policies. The toolkit includes an editor for defining policies and storing them in domains implemented using an LDAP directory service, and a Java-based policy interpreter for obligations. We have prototyped translators from the Ponder language into Windows policy templates and firewall policy rules. A framework was implemented for the enforcement of a previous version of authorisation policies in distributed object systems [21]. We have implemented tools for analysis of policies to determine conflicts between policies specified by different administrators. This detected modality conflicts that may occur for positive and negative policies applying to the same subject and target domains [9].

Further work remains to be done on this framework both for its implementation and for a more efficient and flexible use of its concepts. In particular, we are extending the policy language to cater for other types of policies e.g., delegation, filtering, backing, and looking for a wider use of relationships specified in terms of policies and interaction protocols. Such relationships can be used to describe contractual arrangements between parties and can be enriched with a Quality of Service (QoS) description notation to specify Service Level Agreements (SLA). Our policies can be used to extend traditional SLAs with authorisations and dynamic aspects such as when reports must be provided or what actions the provider must perform if the QoS targets are not reached.

Acknowledgements

The work presented in this paper was funded by the EPSRC under Research Grants GR/L96103 (SecPol), GR/M86109 (Ponds) and GR/L76709 (Slurp), and by Fujitsu Network Systems Laboratories (Pro-Active Role Based Management for Distributed Services Project). We gratefully acknowledge their support.

REFERENCES

[1] BIDDLE, B. and E. THOMAS, Eds. *Role Theory: Concepts and Research*. New York, Robert E. Krieger Publishing Company, 1979.

[2] DAMIANOU, N., N. DULAY, E. LUPU and M. SLOMAN, The Ponder Policy Specification Language, *Proc. Policy 2001: Workshop on Policies for Distributed Systems and Networks*, Bristol, UK, 29–31 Jan. 2001, Springer-Verlag, *LNCS 1995*, pp. 17–28.

[3] DAMIANOU, N., N. DULAY, E. LUPU, and M. SLOMAN. Ponder: A Language for Specifying Security and Management Policies for Distributed Systems. *The Language Specification – Version 2.3. Research Report DoC 2000/1*, Imperial College, Department of Computing, Oct. 2000.

[4] Enterprise-Viewpoint Reference Model, CD 15414, ISO/IEC JTC1/SC7 N2187, 1999.

[5] ETSI. Digital Cellular Telecommunications Systems (Phase 2); (GSM 12.00–12.04, 12.08, 12.20). ETS 300 612–(1–5, 22, 27), 1996.

[6] LUPU, E., N. DULAY, N. DAMIANOU and M. SLOMAN, Ponder: Realising Enterprise Viewpoint Concepts, *Proc. 4th Int. Enterprise Distributed Object Computing (EDOC2000)*, Mukahari, Japan, 25–28 Sept. 2000, IEEE Press, pp 66–75.

[7] LUPU, E. *A Role-Based Framework for Distributed Systems Management*. Ph.D. Dissertation, Imperial College, Dept. of Computing, London, UK, 1998.

[8] LUPU, E.C., and M.S. SLOMAN. Reconciling Role Based Management and Role Based Access Control. *2nd ACM Role Based Access Control Workshop*, Fairfax, VA, 1997, pp. 135–142.

[9] LUPU, E.C., and M. SLOMAN. Conflicts in Policy-Based Distributed Systems Management. *IEEE Transactions on Software Engineering, Special Issue on Inconsistency Management*, 25(6), Nov. 1999, pp. 852–869.

[10] MAGEE J. and J. MOFFETT eds. (1996). *IEE/BCS/IOP Distributed Systems Engineering Journal Special Issue on Services for Managing Distributed Systems*, 3(2), 1996.

[11] Marriott, D.A. and M.S. Sloman. Implementation of a Management Agent for Interpreting Obligation Policy. *IFIP/IEEE Distributed Systems Operations and Management*, L'Aquila, Italy, Oct. 1996.

[12] Moffett, J.D. and E.C. Lupu. The Uses of Hierarchies in Access Control, *4th ACM Role Based Access Control Workshop*, Fairfax, Virginia, Oct. 1999.

[13] Moore B., J. Strassner., E. Elleson., A. Westerinen, *Policy Framework Core Information Model*, IETF draft, Oct. 2000. Available from http://www.ietf.org

[14] Mouly, M. and M.-B. Pautet. *The GSM System for Mobile Communications.* Published by the authors ISBN 2–9507190–0–7, 1992.

[15] OMG Object Management Group, Object Constraint Language Specification, Chapter 7 in *OMG Unified Modelling Language Version 1.3*, June 1999.

[16] Sandhu, R.S. et al. Role-Based Access Control Models. *IEEE Computer*, 29(2):38–47, 1996.

[17] See http://www-dse.doc.ic.ac.uk/policies/ for links to academic and industrial work on policies.

[18] Sloman, M. Policy Driven Management for Distributed Systems. *Journal of Network and Systems Management*, 2(4):333–360, Plenum Press, 1994.

[19] Sloman, M.S. and K.P. Twidle. Domains: A Framework for Structuring Management Policy. Chap. 16 in *Network and Distributed Systems Management*, M. Sloman ed., Addison-Wesley, 1994, pp. 433–453.

[20] Virmani, A., J. Lobo, M. Kohli. Netmon: Network Management for the SARAS Softswitch, *IEEE/IFIP Network Operations and Management Symposium*, (NOMS2000), ed. J. Hong, R. Weihmayer, Hawaii, May 2000, pp. 803–816.

[21] Yialelis, N. and M.S. Sloman. A Security Framework Supporting Domain Based Access Control in Distributed Systems. *ISOC Symposium on Network and Distributed Systems Security*, San Diego, California, Feb. 1996.

Chapter 4

An agent-based framework for large-scale Internet applications

Mamadou Tadiou Kone
Japan Advanced Institute of Science and Technology, Ishikawa-ken, Japan

Tatsuo Nakajima
Waseda University, Tokyo, Japan

1. Introduction

The tremendous growth of the Internet in the past few years sparked a whole new range of applications and services based on its technologies. Users will be able to take full advantage of these new capabilities only if there is an appropriate configuration to deal with the scalability and heterogeneity problems inherent to the Internet. In this line, resource discovery on the network and Quality of Service (QoS) assurance are important subjects that are gaining attention. In particular, the Service Location Protocol (SLP) [PERK 97] designed by the Internet Engineering Task Force (IETF) aims to enable network-based applications to automatically discover the location of services they need. However, SLP was designed for use in networks where the Dynamic Host Configuration Protocol (DHCP) [DROM 97] is available or where multicast is supported at the network layer. Neither DHCP nor multicasting extends to the entire Internet because these protocols must be administered and configured. As a result, SLP does not scale to the Internet.

Our objective in this paper is to deal with two important limitations in resource management for large-scale applications: scalability and communication costs. We propose in this paper a framework that relies on the concepts of multi-agent systems and Agent Communication Language (ACL) described in [PATI 98]. In this framework, a user agent, a QoS manager agent, one or several facilitator agents, and service agents (application agent, system agent, network agent, and resource agent) engage in a mediated communication through the exchange of structured KQML messages. Following this introduction, we state in Section 2 the problem we intend to examine. In Section 3, we describe the concepts and protocols underlying our multi-agent system-based QoS negotiation scheme. In addition, we give the implementation details of our framework in the same section.

Some issues and perspectives are proposed in Section 4 and then related works are presented in Section 5. Finally, we conclude in the last Section 6.

2. Agent-based systems

2.1. Multi-agent systems

There are two well-known perspectives in defining the word *agent*: the software engineering perspective and the cognitive science (AI) perspective both explained in [GENS 94]. The first refers to a piece of software called *mobile agent* or *autonomous agent* that can migrate autonomously inside networks and accomplish tasks on behalf of their owners. On the other hand, the second states that *multi-agent systems* are distributed computing systems composed of several interacting computational entities called agents. These constituent agents have capabilities, provide services, can perceive and act on their environment. Service components involved in a QoS provision are modeled as this type of agent.

2.2. Agent communication languages

Agent communication languages (ACL) introduced in [KONE 00] stem from the need for better problem solving paradigms in distributed computing environments. One of the main objectives of ACL design is to model a suitable framework that allows heterogeneous agents to interact, communicate with meaningful statements that convey information about their environment or knowledge.

The Knowledge Sharing Effort group designed one example of ACL, the Knowledge Query and Manipulation Language (KQML) described in [PATI 98]. Our framework uses the KQML language, which comprises three layers (Figure 1): the communication layer, the message layer, and the content layer. A KQML message has the following structure:

(*tell*

       :sender QoS-manager
       :receiver User
       :language Prolog
       :in-reply-to idl
       :ontology QoS-ontology
       :content "available (resource, URL)"

Communication Layer: sender, receiver, msg id
Message Layer: performatives, msg format
Content Layer: ontology, content language

Figure 1. *KQML three layers structure*

Here, *tell* is called a *performative, :sender, :receiver, :language, :in-reply-to*, and *:ontology* are parameters. The QoS manager informs (*tell*) the user about the availability of a resource on the Internet by using Prolog as a content language. There are two types of agent communication: the direct communication relates a sender agent with a known receiving agent and the mediated communication illustrated in Figure 2 uses the services of special agents (facilitators) that act as brokers between agents in need of some service and other agents that provide them. Mediation involves on one hand, needy agents subscribing to services and on the other hand facilitators brokering, recruiting, and recommending agents that registered their identities and capabilities.

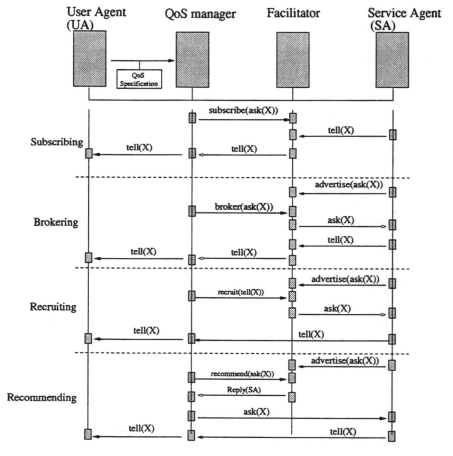

Figure 2. *Facilitator mediated QoS negotiation*

3. Multi-agent system-based QoS negotiation

3.1. The problem

In standard QoS provision schemes for applications running on small or local area networks, a QoS manager determines all configurations that can sustain an activity by:

- identifying necessary system components and building potential configurations;
- classifying these configurations; and
- selecting the most suitable configuration.

This approach assumes that the QoS manager has knowledge of potential service providers, system components and resources that exist in its environment and can communicate directly with them. As long as the number of entities involved in this service is small, this scheme is feasible and communication costs are acceptable. However, in a heterogeneous setting like the Internet with millions of computers, this approach shows two clear limitations:

- *First*: During negotiation, the QoS manager alone must bear all the burden of identifying and selecting appropriate resources on a large-scale network like the Internet. This situation adds a substantial overload on the operation of the QoS manager. In addition, services and resources may not be guaranteed consistently.
- *Second*: When the number of entities involved in a direct communication with the QoS manager is modest, communication costs remain reasonable. However, in the Internet, these costs become prohibitive even with auxiliary local QoS managers.

To address these scalability and communication costs issues, we propose a framework for QoS negotiation illustrated in Figure 3 where client applications and service providers engage in a mediated communication. The mediators called *facilitators* and *QoS brokers* are supplied with information about identities and capabilities of service providers by the providers themselves. These entities are modeled as software agents with attributes, capabilities and mental attitudes as in AI. At the core of our framework lies the concept of a multi-agent system composed of a user agent, a QoS manager agent, a facilitator agent, and service agents (network agents) communicating in KQML.

3.2. Concepts and framework description

3.2.1. Concepts

Prior to starting a service, a user specifies and supplies the QoS manager with a level of service expressed in QoS parameters. Then, the QoS manager must identify the set of components that can sustain this service. This process uses the following concepts:

- An ontology provides a vocabulary for representing and conveying knowledge

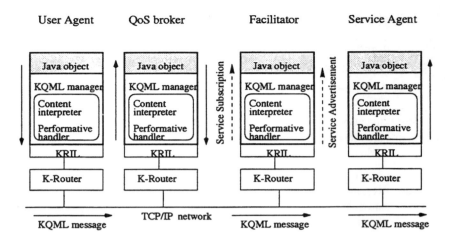

KRIL : KQML Router Interface Language

Figure 3. *System architecture*

about a topic (e.g. QoS) and a set of relationships that holds among the terms in that vocabulary. Our architecture uses four ontologies:

  * a *yellow page* ontology for service advertisement by service agents;
  * a *white page* ontology for finding the location of an agent given its name;
  * a *general QoS* ontology for the current domain knowledge;
  * and a *QoS broker* ontology for asking network options by the user and QoS broker.

● A *KQML manager* encompasses:

  * *Conversations* that group messages with a common thread identified by the ":reply-with and :in-reply-to" parameters;
  * *content interpreters* that handle incoming and related response messages according to the ACL, content language and ontology associated to these messages;
  * *performative handlers* that process a message performative in conjunction with its ACL, content language and ontology.

3.2.2. QoS negotiation protocol

In our framework, four types of agents communicate in KQML according to the following protocol:

● The user informs its agent via an interface of the required level of service.
● The user agent sends to the QoS manager agent a KQML message with required levels of service expressed in appropriate QoS parameters like in Figure 4.

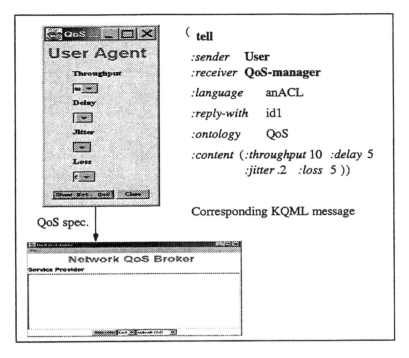

Figure 4. *User and QoS broker interaction*

- The QoS manager needs to identify all components necessary to build a configuration that can sustain an activity. For this purpose, its agent sends a KQML message to the facilitator agent and can ask its cooperation in four different ways (subscription, brokering, recruiting and recommendation) in discovering all the appropriate resources. A structure of this KQML message and agent interaction is shown in Figure 5.
- The facilitator agent acts as a resource broker that:
  * recruits, recommends appropriate service agents (application, system, and network agents) to the QoS manager;
  * forwards the QoS manager messages (brokering and recruiting) to suitable service agents; and
  * informs (on subscription) or recommends to the QoS manager service agents that fulfill its requirements.
- All service agents (network agents) advertise their capabilities to the facilitator agent upon registration. Upon request from the QoS broker, the facilitator supplies the identities and locations of necessary network resources. At last, the user may view on an appropriate interface the available resources.

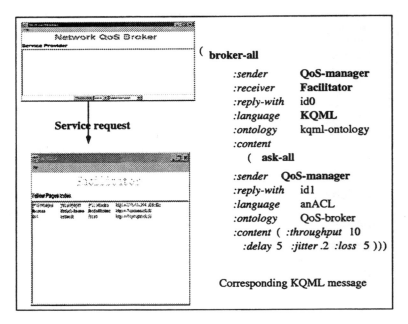

Figure 5. *QoS broker and facilitator interaction*

This QoS negotiation model for large-scale Internet applications is applied in two ways: locally or remotely. When the required resources are available locally and registered at the local facilitator, negotiation is done at the current host as illustrated in Figure 6. On the other hand, when some resources are unavailable on site, the local facilitator reaches out to other facilitators at different locations as illustrated in Figure 7. The local facilitator forwards requests (*broker-all*) to remote facilitators which in turn conduct a local inquiry. In fact, this approach to agents' interaction is already used in the field of *agent-based software engineering* where application programs are modeled as software agents and interaction is supported by an appropriate ACL. In this approach, agents are organized in a *federated system* with messages relayed by facilitators between hosts.

3.3. Implementation

In experimenting with this model of resource discovery and QoS negotiation, we designed a prototype in the JAVA language to simulate QoS negotiation between several agents at the network level. That is to say, to illustrate our approach, a user agent and network agent communicate via a QoS broker and a facilitator in terms of network parameters only. First, local negotiation is considered, then it is extended to remote locations across the Internet.

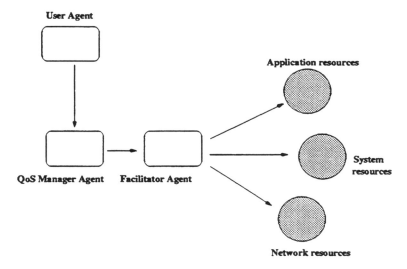

Figure 6. *Local QoS negotiation with a single facilitator dealing with resources inside a given host*

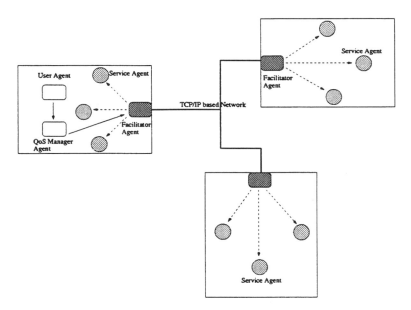

Figure 7. *Large-scale QoS negotiation with several facilitators involved in the negotiation process across the Internet*

The implementation of our prototype includes the following tools:

• The Java-based KQML API called *JKQML* in [HAJI 98]. The JKQML API with its structure in Figure 8 adapted from [HAJI 98] provides a platform for designing KQML-enabled agents. JKQML is based on the JAVA language and provides interoperability to software that needs to exchange information and services.

Handling KQML messages involves the following steps:

1. Instantiating a KQML manager with the method:
   public KQMLManager(String agentName, String protocol, int port);
2. Managing protocol handlers with the method:
   public void addProtocol(String protocol, int port);
3. Managing content interpreters with the method:
   public void addContentInterpreter(String acl, String language, String ontology);
4. Managing performative handlers with the method:
   public void addPerformativeHandler(String acl, String language, String ontology, String performative, PerformativeHandler ph);
5. Managing conversation termination with the method:
   public void setConvCleanupHandler(ConvCleanupHandler c).

• We used the Stanford KSL *Ontolingua* ontology editor at [GRUB 92] to design both the general QoS ontology and the QoS-broker ontology used by the

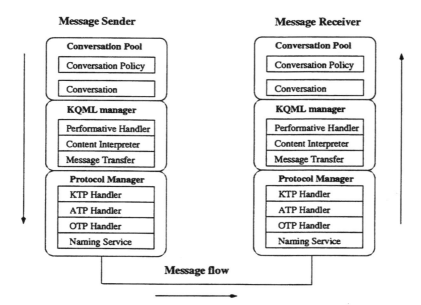

Figure 8. Structure of JKQML

language interpreter. Then, we extended our simulation program to a larger TCP/IP network with facilitators at different locations communicating in KQML. A couple of networks with different characteristics (throughput, delay, jitter, and loss) were discovered successfully and displayed on a local user interface. Figures 4 and 5 illustrate some transactions between the participating agents.

4. Issues and perspectives

In an open and heterogeneous environment like the Internet, agents that interact and coordinate their activities face some major challenges:

– How can they find one another and especially, locate the facilitators? As the number of facilitators grows, finding their location becomes a real concern. The idea of introducing a *facilitator directory* that forwards external inquiry to all facilitators across the Internet could address this problem.
– Although many ACLs exist today, the communication language chosen should express concisely the message content of an agent. That is to say, the message semantics of an ACL must be consistent across platforms.
– With any kind of message transport protocol (KQML transport protocol (*ktp*) or agent transport protocol (*atp*)), the issue of fault tolerance due to network failure remains. Multi-agent systems must rely on a robust and reliable environment. However, the heterogeneous nature of the Internet offers no guarantee.

In addition to negotiation on the network layer, we are looking forward to extending our model to the application and system layers as well. This way, with a suitable QoS translation scheme between these layers, it is possible to cover a complete end-to-end QoS negotiation.

We intend to investigate the alternative of mobile agents as a message transport protocol. Enabling facilitators to move around the network, deliver information and collect advertisements like mobile agents is an option we are interested in. These *mobile facilitators* can interact on site with local QoS brokers and service agents. In addition, as the new Foundation for Intelligent Physical Agents (*FIPA*) ACL standard is emerging, we are looking forward to implement our model in this language.

5. Related work

A number of service discovery protocols have been implemented for different platforms. Some examples are the Service Location Protocol (SLP) designed by the Internet Engineering Task Force, the Dynamic Host Configuration Protocol (DHCP), the CORBA architecture [OTTE 96] with its *Trader* and *Naming Services*, and recently Sun Microsystems's Jini.

5.1. The Service Location Protocol (SLP)

The idea of using multiple agents for the discovery of services across a local area network has already been used by the SLP. In this model, a user agent (UA) acts on behalf of a user or client in need of a service while a service agent (SA) declares its services to a directory agent previously discovered. In addition, a directory agent (DA) accepts requests and registrations from a UA or a SA.

There are two fundamental differences between the SLP scheme and our approach: SLP uses multicast and DHCP protocols to initialize its scalable service discovery framework. However, as DHCP cannot extend to the entire Internet, SLP is unable to scale to the entire Internet. A user agent itself must send its queries to a remote DA when a service is not available locally. In contrast, our approach considers a federation of services as illustrated in Figure 7 with several facilitators. Only facilitators may forward requests from one region to another. In addition, we use KQML messages to convey these requests across the Internet.

5.2. The CORBA Trader and Naming Services

CORBA is a middleware that enables a *client application* to request information from an *object implementation* at the server side. In addition, CORBA can advertise available objects and services on behalf of object implementations via a *Common Object Services Specifications* (COSS) service called *the Trader Service*. Services are registered with the *Naming Service* by specifying its name and object reference. A client who wishes to access the service specifies the name of the service, which the Naming Service uses to retrieve the corresponding object reference, whereas services are registred with the Trader Service by specifying its service type, properties and object reference. A client who wishes to access the service, specifies the type of the service and constraints. Therefore, the Trader Service can be viewed as a yellow pages phone book.

In spite of the similarities in both approaches, it is important to note that the main difference between our system and CORBA services is that we are dealing with messages which bear meaning and are organized in conversations. The players in our system are agents that are engaged in structured conversations while CORBA enables applications to exchange only objects, data structures, and propositions.

5.3. Jini

Jini presented in [EDW 99] is a network operating system by Sun Microsystems aimed at a broad range of electronic devices and software services assembled in a single distributed computing space. Although the components work together to serve a common goal, they are still identified as separate components on a network. The Jini discovery architecture is similar to that of SLP. Jini agents discover the existence of a Jini look-up server, which collects service advertisements like the facilitators in our system. Jini agents then request services on behalf of client softwares by contacting the look-up server.

6. Conclusion

In this paper, we have presented a framework for resource discovery and quality of service negotiation over the Internet. The main point is that our framework relies on the concept of multi-agent systems and agent communication language. In contrast to automatic resource discovery protocols like the SLP, our scheme scales to the entire Internet. To illustrate its effectiveness, we designed a prototype based on the IBM Java KQML API with several agents: user agent, QoS broker agent, facilitator agent, and network agents that interact in the KQML agent communication language. Although this approach may look attractive, its main drawback lies in the important number of facilitator agents that the system must deal with. In the future, we intend to let these facilitators move from host to host with information just like mobile agents.

REFERENCES

[EDW 99] Keith E. Edwards, *Core Jini*, Prentice Hall, Upper Saddle River, New Jersey, 1999.

[GENS 94] Michael R. Genesereth, Steven P. Ketchpel, "Software Agents", *Communications of the ACM*, vol. 37, num. 7, 1994, p. 48–53.

[GRUB 92] Thomas R. Gruber, *"Ontolingua: A Mechanism to Support Portable Ontologies"*, 1992.

[HAJI 98] Hajime Tsuchitani, Osamu Furusawa, *"JKQML"*, 1998, AlphaWorks, IBM.

[KONE 00] Mamadou T. Kone, Akira Shimazu, Tatsuo Nakajima, "The State of the Art in Agent Communication Languages", *Knowledge and Information Systems (KAIS)*, vol. 2, num. 3, 2000, p. 1–26.

[OTTE 96] Randy Otte, Paul Patrick, Marc Roy, *Understanding CORBA. The Common Object Request Broker Architecture*, Prentice Hall, Upper Saddle River, New Jersey, 1996.

[PERK 97] C. Perkins, "Service Location Protocol", *White paper*, May 1997, Sun Microsystems.

[DROM 97] R. Droms, "Dynamic Host Configuration Protocol", *Technical report num. rfc2131*, Oct. 1997, IETF, Network Working Group.

[PATI 98] Ramesh S. Patil, Fikes E. Richard, *"The DARPA Knowledge Sharing Effort: Progress Report"*, p. 243–254, Morgan Kaufmann, San Francisco, California, 1998.

Chapter 5

An agent-based paradigm for managing service quality in open network environments

Alan Marshall

Advanced Telecommunications Systems Laboratory, School of Electrical and Electronic Engineering, Queen's University of Belfast, Northern Ireland, UK

Nazim Agoulmine

LIP6 Laboratory, University of Paris VI, France

1. Introduction

The intermediate nodes (e.g. routers) in the present day Internet are mostly vertically integrated closed systems whose functions are rigidly programmed into the embedded software and hardware by the vendor. Programmability is limited and this results in slow introduction of new services, slow standardisation process, and redundant operations across the protocol layers. The paradigm "Programmable Network" envisions the future telecommunications network as an open system that can be programmed to deliver QoS-based network services (e.g., voice, video and data globally [1]). Quality of Service (QoS) represents the set of requirements, imposed by the user, related to the performance to be achieved in order to support the provision of a particular service. Resources such as network bandwidth, processing time and memory are still scarce inside and outside the network and need to be controlled in very tight manner. Active packets or open signalling have been proposed as the mechanism to achieve this programmability.

However, it is important to provide a user with mechanisms that permit one to express this QoS. Service Level Agreement is the formal negotiated agreement between a service provider and a customer for service delivery. It is designed to create a common understanding about services, priorities, responsibilities, etc. SLAs can cover many aspects of the relationship between the customer and the service provider such as quality of services, customer care, billing, provisioning, etc. SLA is at the heart of service quality management in open network environments.

Agent technology has been widely proposed as a key technology in moving towards an open, active and programmable network environment [2]. Agents enhance the autonomy, intelligence, and mobility of software objects and allow them to perform collective and distributed tasks across the network. Agents may be dispatched to the nodes spanning the routes in a network, and will be responsible for the maintenance of services through the virtual private networks (VPN) created. Maintenance of a VPN may involve dynamic reconfiguration, rerouting of connections, and renegotiation of QoS targets.

Scheduling of different classes of traffic within the switches and routers has been identified as one of the most important resources to be reconfigured in order to support QoS. There has already been substantial research done within an area of re-configurable ATM switches [12, 13, 14]. The adaptive switch approach allows the reconfiguration of queuing scheme based on traffic profile, while maintaining a certain level of QoS for different classes of traffic. As a further step for the reconfiguration of scheduling mechanisms to modify the queuing strategy in routers, a mechanism called, *Active Scheduling* has been proposed [15, 16]. Active Scheduling allows the introduction of a procedure by which the functionality of a router can be dynamically modified. Such adaptation may be initiated by user applications, third party service providers, or by a network operator. In this scheme we dynamically alter the queuing system of heavily loaded routers, according to different classes of traffic (voice, video and data), since each class has different traffic characteristics. Dynamic adaptation of the scheduling policy according to either traffic load or user demand offers the potential to implement prioritized fair queuing and service class guarantees based on traffic load profile and on negotiated service parameters such as delay and price. This is achieved by sending an active packet to the router initiating the reconfiguration of queuing and scheduling policies.

A number of specific enabling technologies have been identified that will allow the introduction of programmability into the proposed framework. These technologies are: *Active Networks, Agents, Middleware*, and *Dynamically Reconfigurable Hardware*. The developed framework also proposes different levels for the implementation of these technologies within the network.

The remainder of the paper is organised as follows: Section 2 describes the Quality of Service requirements, Section 3 presents enabling technologies of active packets and agents, Section 4 presents a framework of adaptive network, Section 5 describes active scheduling mechanism and active router architecture. We then discuss future work and give concluding remarks.

2. Quality of Service

Quality of Service (QoS) is the keyword used to represent the set of requirements, imposed by the user, related to the performance to be achieved in order to support the provision of an application. Parameterization of QoS parameters is an important research subject that is carried on by a number of research teams and some results are already available [8, 30]. However, the impact of the QoS variation on the application and the end user perception is still an open issue. In distributed multimedia systems, the desired Quality of Service has to be conveyed to the QoS mechanisms in the form of QoS parameters. Parameters can be expressed in very different formats, varying according to the application field, the required level of abstraction, or even as user needs. Two main abstraction levels can be identified in systems subject to QoS specification: the *application level* and the *system level*. An application parameter is generally related to an idea present only at this level (e.g., shown frame number per second of a video broadcast application). At system level, this corresponds to the required network bandwidth to transfer data, delay and delay variation (jitter), the processing time to compress and decompress the information, the memory size, the priority to specialized hardware, etc.

The parameters identified at the application level are presented in Table 1.

At the network level, there is also a set of parameters that identify the quality of the communication service (see Table 2). The parameters are related to the end-to-end transmission of data in the physical network.

Table 1. Generic QoS application parameters

Parameters	Value range	Descriptions
App::[Data Unit Size]	Min, Max	Date unit size processed by the application (bits)
App::[Data Unit Rate]	Min, Max	Data unit rate required from the application (units/s)
App::[End to End Delay]	Min, Max	End-to-end data unit transfer time supported by the application (ms)
App::[End to End Jitter]	Min, Max	End-to-end data unit transfer delay time variation supported by the application (ms)
App::[Error Ratio]	Max	Error ratio supported by the application (%)

Table 2. Generic network parameters

Parameters	Value range	Descriptions
Net::[Bandwidth]	Min, Max	Bandwidth allowed by the underlying network (bits/s)
Net::[Delay]	Min, Max	End-to-end delay time for the transmission of data unit in the network
Net::[Loss]	Min, Max	Loss of data unit in the network

At the system level, we have identified two main parameters (this list is not exhaustive and can be extended in the future): the CPU time allowed to the application and the size of RAM to be reserved are presented in Table 3.

Table 3. Generic OS parameters

Parameters	Value range	Descriptions
OS::[CPUTime]	Min, Max	% of processing time reserved to the application in the particular OS
OS::[Memory]	Min, Max	Random memory size allowed to a particular application in the particular OS

After the specification phase, the QoS parameters have to be translated between different levels of abstraction to be meaningful for the mechanisms present at the following level. Translation implies that a mapping exists between parameters at different levels. Mappings are not usually one-to-one between parameters, but may be one-to-many, many-to-one or many-to-many parameters [28]. In addition, mappings and translation mechanisms have to be bi-directional, to be able to transfer QoS data from the bottom layers to the top ones, reporting QoS measures to the application, using QoS parameters comprehensible at this level of abstraction.

In order to achieve the desired system performance, QoS mechanisms have to guarantee the availability of the shared resources needed to perform the services requested by users. The concept of resource reservation provides the predictable system behavior necessary for applications with quality of service (QoS) constraints. The development of ATM has been a significant advance towards the provision of QoS-constrained communication services. Aiming to provide similar behavior, but working at the logical network level, the IETF is proposing a new suite of protocols for the Internet [8]. The evolution of IP networks from their current best-effort delivery of datagrams into guaranteed delivery of time-sensitive services is still far from complete. Resource Reservation (RSVP) [19] and Differentiated Services (Diffserv) [20] have been proposed as mechanisms for implementing service quality in IP-based networks. However, RSVP has problems with scaling and its implementation on nodes that transport a large number of traffic streams is impractical. Alternatively, Diffserv provides a granular service, which cannot provide fully optimised network performance. It is now considered that future networks will consist of combinations of these mechanisms: RSVP may be more efficiently employed in the access network, whereas Diffserv will be deployed in the higher density backbone [21].

Resource reservation combined with adaptation entails a more relaxed approach for providing QoS to applications. According to this new approach, resources can be seen by applications as guaranteed during some time, but their availability can vary over long periods. Applications are responsible for estimating their initial resource requirements and for negotiating their reservation with resource providers. During run time, the application should be able to adapt its behavior according to QoS shortage notification from the QoS management architecture. QoS mechanisms have to be aware of the possibility of resource adaptation, making it transparent to the application whenever possible [13, 18]. When the agreed QoS is not reachable with the resources available, the application has to be informed in real time that the agreed QoS has to be renegotiated. Applications have to be ready to handle this kind of situation, without severe disruptions in the service being provided to the user. More specifically, applications holding resources, subject to changes in their availability have to be able to degrade gracefully when it occurs. This means that the multimedia applications should be designed in such a way that they can handle various situations. This makes them portable over various systems and network technologies without any redesign. This implies a strong definition of the semantic interaction between the application and the QoS management architecture through a well-defined interface. This interface should be flexible enough to cope with the various underlying technologies.

3. Enabling technologies for adaptation

3.1. Active/programmable networks

There are currently two favoured approaches for introducing programmability into the network: OPENSIG (Open Signalling) and Active Packets [3, 4, 5, 6]. The OPENSIG community is of the view that by modelling communication hardware using a set of open programmable network interfaces, open access to the switches and routers can be provided. These open interfaces allow service providers to manage and construct new network services by using distributed system middleware toolkits, e.g., CORBA [5].

Active packets allow customised computation on the users' data within the intermediate nodes (e.g., routers). The user of an active network sends a customised program to a node in the form of an active packet and requests the node to process that program. These networks are active in two ways [5, 6, 7, 8]: Routers perform computations on the user data flowing through them; USERS can inject programs into the network, thereby tailoring the node processing to be user and application specific network encapsulation protocol [9]. The routers that perform computations of user active packets are called active network node. They provide a secure environment for running mobile programs of high-level languages or active packets on a low layer, e.g., IP, as such an integration will facilitate service logic customisation, rapid service creation, service migration and integration. Obviously it is the highly available programmability of active networks that allows a promising solution for meeting the challenging demands such as flexible and dynamic service provisioning and controlling.

Although there is as yet no active network standard, researchers are currently investigating signalling, control and management aspects. The IEEE's P1520 sub-networking group is the main focus for the creation of a reference model for open and programmable networks [1]. Research in active/programmable networks can be broadly classified as:

– Programmability of control architectures, i.e., creation of control architectures by operators and thus creation of virtual private networks (VPNs) [10].

– Creation of multiple coexisting execution environments through appropriate operating system support and active network encapsulation protocol [11].

3.2. Agent technologies

The agent concept has been widely proposed and adopted within both the telecommunications and Internet communities. It is a key tool in the creation of an open, heterogeneous and programmable network environment. This trend is motivated by the desire to use the agents to solve some of the problems encountered in large scale distributed and real-time systems such as the volume and complexity of the tasks, latency, delays, and others. Generally, an agent can be regarded as an assistant or helper, which performs routine and complicated tasks on the user's behalf. In the context of distributed computing, an agent is an autonomous software component that acts asynchronously on the user's behalf. Agent types can be broadly categorized as static or mobile.

The majority of current communication system architectures employ the client-server method that requires multiple transactions before a given task can be accomplished. This can lead to increased signalling traffic throughout the network. This problem can rapidly escalate in an open network environment that spans multiple domains. As an alternative solution, mobile agents can migrate the computations or interactions to the remote host by moving the execution there. For example, mobile agents can be delegated to complete specific tasks on their own, providing that a certain set of constraints or rules have been defined for them [17]. They can then be dispatched across the network in the form of mobile program or mobile code that can be recompiled and executed in the remote host.

The main motivation of the use of agent technology in this work is driven by the desire to automate the control and management processes by allowing for more programmability of the network to rapidly customize the provision of new information and telecommunication services. Hence, agents can also be used to implement Service Level Agreements (SLAs) between different actors of the network and service era. Agents can then be used as brokers or mediators between end users and a service provider in order to implement the SLA. In this way, complicated QoS metrics (from end user's point of view) can be communicated in a simplified manner. Service provider and network provider agents can then negotiate with users' agents in order to meet the required service [18].

3.3. Middleware technology

Network applications communicate with one another in order to deliver high-level services such as controlling a multimedia communication session or managing a set of physical or data resources. The creation of QoS-based communication services, and management of network resources can be conflicting objectives that network designers have to meet. A logical communication mechanism among network entities, which hides the underlying implementation details of hardware and software platforms, is the most important function of middleware technology that can help realize the network designer's objectives.

Object Management Group CORBA (Common Object Request Broker Architecture) is currently one of the most popular middleware technologies. This is due to its interoperability, independence of platform, operating system, programming language, type of network and protocols. Based on CORBA, the OMG has provided a set of specifications that permits the interoperability between agents called MASIF (Mobile Agent System Interoperability Facility) [27]. In the proposed framework, MASIF-CORBA is used as the middleware between service layer and the network layer (Figure 1). The CORBA middleware provides a communication interface for different agents while MASIF provides the facilities for agent transfer, management and naming.

3.4. Dynamically reconfigurable hardware

Programmability and reconfigurability have been used extensively in the communications industry. There are now a number of systems that exploit reconfiguration in order to adapt their system (including hardware) resources to meet changing requirements. Reconfigurability has been incorporated into products in order to overcome issues resulting from a lack of standards, or the slowness of the standardization process. The ITU V.90 [23] modem standard is one of the more recent examples, being sold on the promise that it could be upgraded. However, while static reconfiguration of a system's hardware resources is now successfully employed, the concept of dynamically reconfiguring those resources is still a matter for further research. Dynamic reconfiguration implies that the hardware may be reconfigured during run-time. This can be achieved using Dynamic Reconfiguration Logic (DRL) technology such as may be found in the more recent FPGA families, for example the Virtex series family by Xilinx [24] which can be reconfigured within microseconds and permits partial reconfiguration of some part of the device whilst another part is still operating. This permits the concept of changing or reconfiguring the hardware in real-time as the data is being processed [25].

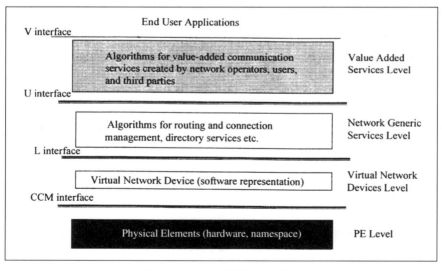

Figure 1. Overview of the IEEE's P1520 architecture

4. A framework for open adaptive architecture

The current Internet service model is a "best-effort" datagram service. However current developments are changing this situation dramatically: the future Internet is envisioned as a multi-service platform, supporting real-time traffic and complex multi-user services. The current Internet model is currently in the process of being extended to include quality of service (QoS) guarantees. Due to the inherent time-sensitive characteristics of many multimedia applications (e.g., audio and video conferencing, multimedia information retrieval, etc) the network is required to provide a wide range of QoS guarantees (with respect to bandwidth, packet delay, delay jitter and loss). The guaranteed bandwidth must be sufficient to accommodate video and audio streams at acceptable resolutions, while the end-to-end delay must be small enough for interactive communication. In order to avoid breaks in audio and video streams' playback, delay jitter and loss must be sufficiently small. Over-allocation of bandwidth reduces link (and networks) utilisation, and this may be a critical factor in architectures that support third party Virtual Private Networks (VPNs). Hence, the diversity of traffic characteristics and performance requirements of existing applications, as well as uncertainty about the future applications, requires the scheduling discipline in the network elements (i.e., routers) to allocate fair delay, bandwidth and loss rate guarantees [RENV].

The two favoured architectures for providing QoS for time-sensitive services over IP networks are the Integrated Services (IntServ) [19] and Differentiated Services (DiffServ) [20]. IntServ requires applications to signal their service requirements to the network via reservation requests. These take the form of the Resource Reservation Protocol (RSVP). DiffServ works in the core network and employs a scalable aggregation to provide prioritisation among classes of traffic. While RSVP has problems with scaling to large networks (where a large number of nodes need to maintain soft-state information on all traffic flows), DiffServ provides only a granular service which cannot provide fully optimised network (or application) performance. It is now considered that future networks will consist of combinations of these mechanisms: RSVP may be more efficiently employed in the access network, whereas Diffserv will be deployed in the higher density backbone [21].

The research described here develops a new framework for the dynamic adaptation of networks using programmable networking paradigms coupled with the use of mobile and intelligent agent technologies. Such networks will be heterogeneous environments consisting of more than one network operator. In this environment, agents will act on behalf of users or third party operators to obtain the best service for their clients. In the framework agents may be dispatched to nodes within a network and will be responsible for the maintenance of services through the virtual private networks (VPNs) created. Maintenance of a VPN may involve dynamic rerouting of connections and renegotiation of QoS targets in nodes. An important feature of the research proposed here is the mapping of higher layer requirements onto actual physical resources in the network equipment. The renegotiation of QoS can be achieved through modification of scheduling schemes in routers and switches spanning the VPN. The framework is based on the emerging standard for open interfaces for programmable network equipment, namely the IEEE's P1520 reference model. To date, most of the research in this area has tended to concentrate on the network generic services layer (between the U and L interfaces). This architecture has a base principle: a clear separation of switching/routing functions from network service control and management tasks through the U interface while the L interface provides an abstraction from the network device architecture. Such a separation brings a great flexibility to active networks' support for service provisioning when it is enforced in a form of standardised network programming interface.

Because of the diversity in applications, the complexity of implementation and security issues, it is likely that the centralised network management approach will be replaced by a new distributed management system. We call this management architecture "Active Distributed Management Architecture" (ADMA). It can avoid scalability problems and offers flexibility to users, third party operators and network operators. The ADMA uses all the enabling technologies mentioned. This management architecture operates at the network and element levels. The open framework has the following components:

Agents: are used mainly as autonomous negotiators. The *user agents* are used to obtain a connection from a service provider. The Service Level Agreement (SLA) contains the negotiated parameters such as the best price, delays (related to a particular scheduling scheme), guarantee of network availability, backups for the connections. The *service broker agent* acts as negotiator between the clients and the service provider. A *network agent* from the service provider negotiates with different network operators to set up a VPN.

Active network manager: Local programmability of the active node functions and management of resources are provided by means of a mobile agent environment, which enables asynchronous implementations of network-wide services such as policy-based control and management. The Active Network Manager is part of the network operations and contains policy management, accounting, security, and auditing and repository of various services. The policy server within an active manager is responsible for collecting all of this information relevant to each VPN. It then makes decisions based on the SLAs and communicates these decisions to the element layer. The goal of the policy server is to develop a response consistent with the SLA. The response is then transmitted to the network element, e.g., router using active packets. The Active Network Manager sends active packets for the reconfiguration of those routers within its domain. Active packets are chosen instead of conventional signalling methods such as SS7. This is primarily because conventional signalling schemes do not have the method of transporting individual local agents to the active routers, where they operate on behalf of each user agent.

Active packets: Contain codes that can be compiled at the active router, e.g., java codes that can be compiled on the fly and executed in the JVM on router's OS. It can also trigger the activation of a user agent.

Active router: This performs tasks such as dynamic selection and modification of packet scheduling schemes in order to maintain QoS targets, along with other route configuration and maintenance tasks. The active routers' operating system must also support the local agents that have been dispatched from the Active Network Manager, and a manager for them (local agent manager). These components are described below.

Local agent manager: Performs local routine control/management functions. It coordinates data transfer to and from different agents from different sources and clears the information data (duration, packet loss, etc) of a session, when the session ends. While a session is active, the local manager monitors resource usage and feeds back the VPN status information to the ADMA. It also performs local data filtering for security purposes.

Local agents: Are agents acting on behalf of each user, or a third party service provider. Agents should be as light weight (compact size) as possible to enable their real-time installation in the routers.

It may be observed that the network, in addition to the active routers, will also contain legacy routers, and these will still depend on the traditional client-server way of provisioning VPNs. It is highly likely that legacy routers will not be able to support QoS targets, and in this case it is the responsibility of the Active Network Manager, in co-operation with the Active Routers, to set up specific paths for QoS-sensitive traffic flows through the network. It may also be possible for two communicating active routers spanning a legacy router to *compensate* QoS traffic flows through the legacy router, for example by dynamic buffering and prioritization of certain traffic classes. Figure 3 shows a typical QoS mapping scenario at three levels of the architecture. The three levels of QoS attributes correspond to three levels of the architecture in Figure 2.

5. Active router

At the element level, an active router architecture is proposed. The reconfigurable resources at the element level include queuing, scheduling and traffic shaping tools. The active router reconfigures the scheduling schemes. The router serves as a means to implement the options of different classes of service according to the price and delay requirements of the clients. Active packets are used as the mechanism to change the scheduling scheme within the router. An active packet could serve as a trigger to invoke an agent on the router or it can contain the actual code

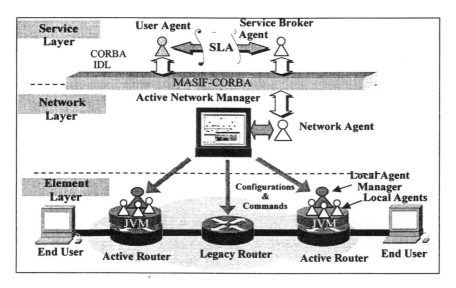

Figure 2. *Architecture of the adaptive network*

User Level		Network Level (End to End)		Element Level (Hop to Hop)
Quality		Bandwidth Boundary for the type of service(price)		Operation System Resources (CPU)
Type of Applications		Latency Limit		Scheduling Schemes
Cost		Jitter		Traffic flow weight
Availability Guarantee (Data) Security		Loss/Error Rate		Queue Length/Delay
Allow changes?		Re-Negotiation/Re-Configuration of SLA/QoS Policy		Re-Configurable/ Programmable?

(QoS Mapping (Application Level) between User Level and Network Level; QoS Mapping (Network Level) between Network Level and Element Level)

Figure 3. *QoS mapping (generic attributes)*

for reconfiguration. This scheme is called *Active Scheduling* [15, 16]. The agents are dispatched to the active router from the Active Network Manager. They are sent via Active Network Manager because of the security issues involved. The agents are then executed on the MASIF-like middleware in the router. In this scheme, a MASIF-like middleware based on a JVM is proposed as the environment for agent execution. Based on the SLA, the active manager (within each network operator) assigns a scheduling scheme to the users' VPN. This is then deployed to each active router in the VPN and operated by local agents. The SLA for element level QoS contains the options of price and delays of the queuing strategy. In order to have a standard interface between the agents deployed on the router and the embedded hardware and software, an ORB (Object Request Broker) has been used between the JVM and the Kernel. The reason for using an ORB is to provide on one hand all the support for agent management and mobility and on the other hand a standard L interface to interact with router kernel, since there is a wide variety of hardware and software within routers from different vendors.

The router architecture shown in Figure 4 incorporates a number of reconfigurable scheduling disciplines including: First-in, First-out (FIFO) queuing, Weighted Fair Queuing (WFQ), and Jitter Earliest Due Date (J-EDD) [29]. The scheduling schemes are implemented in hardware, using run-time reconfigurable FPGAs. These FPGAs allow fast reconfiguration of scheduling

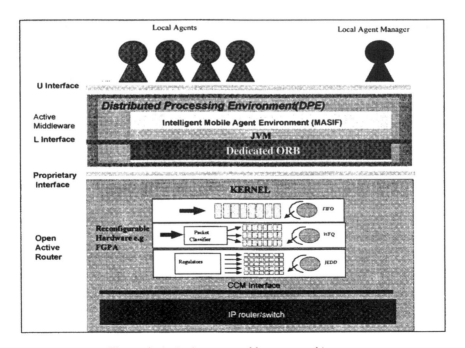

Figure 4. Active/programmable router architecture

schemes on the fly without stopping the scheduler, by downloading the code from the agent to specific sections of the FPGAs while the remaining sections are still functioning. Deciding which type of scheduling scheme is appropriate depends on a number of factors:

Traffic profile: The traffic profiles of different classes of traffic (e.g., video, voice and data) will vary during hours of the day. Reconfiguration of scheduling schemes can take place, according to the traffic profile, for example FIFO is suitable for best effort services (e.g., file transfer, e-mail traffic), while WFQ can be used to implement differentiated or prioritised services for voice and video traffic. If there is a request from some users for a high quality interactive video conferencing, then after negotiation of price, the scheduling scheme is reconfigured to WFQ or J-EDD.

Price issues: Price can vary according to a network operator's tariffing policy, and according to the competition between different operators. We may also have differentiated tariffs (e.g., price of different services is different, due to bandwidth and QoS targets). The scheduling policy can be modified according to the price. High price users can have J-EDD or WFQ for example for video conferencing.

Network congestion: Network congestion will vary over the time of day, and also due to tariff changes. During congestion hours the scheduling scheme for real-time traffic can be changed from J-EDD to WFQ, since J-EDD allocates the bandwidth at the peak rate with no statistical multiplexing. However, under less congested conditions, J-EDD performs better than WFQ and is also preferred for high QoS video applications due to its better performance in handling jitter.

Implementation complexity: The WFQ and J-EDD have same implementation complexity, except that J-EDD does not need round number computation. Prices issues can also be related to the scheduling schemes, keeping in view their implementation complexity. This factor is important for hardware resource usage in the router.

QoS adaptation: As well as selecting different scheduling schemes, it may also be necessary to modify an existing scheme, for example the weights allocated to a traffic flow in a WFQ scheduler may be dynamically changed. In this case, intelligent agents in the routers dynamically reconfigure the weights of their associated services. The agent monitors the accumulated queuing delay for each service. It then reconfigures the scheduler weights (on-the-fly) whenever there is an increase or decrease in the queuing delay for the session. The mechanism defines a window (upper and lower limits) of weights to periodically recalibrate the weights based on queuing delay limits for the per-hop network element. This avoids any short-term unfairness within that frame of delay limits. It offers both dynamic control of QoS, and a high degree of flexibility that is increasingly required in heterogeneous networks where a high variability in the applications' requirements must be supported.

Figure 5 shows dynamic adjustment of weights for two sessions in an active router. The delegated agents adjust the weights whenever there is an increase or decrease in the queuing delay for their sessions. In Figure 5 the queuing delay for session A is maintained between 2ms and 0.7ms. The weight window for session A (continuous traffic flow) is between 40 and 80. The queuing delay for session B (bursty traffic flow) is maintained between 10ms and 1ms. The weight window for session B is between 40 and 80. The weights for sessions are recalibrated by the agent between these limits, in order to maintain the delay limit requirements. The figure shows how each session can be dynamically allocated additional bandwidth up to predefined limits that have been negotiated by the agents.

All the QoS violations can not be recovered only by a local adjustment by a delegated agent. Thus, a delegated agent can also communicate with each other in order to make a load balancing of resources between two active routers. In order to cooperate, delegated agents use a QoS communication protocol. This involves exchanging messages by means of agent services. Three services have been defined in order to support the interactions:

   – **Request_QoS:** The local agent, which has detected a violation in a specific link regarding the initial reservation parameters for the application traffic profile, issues this message to inform a neighbor local agent about shortage. The message contains the level of QoS required to recover from the violation.

   – **Available_QoS:** This message is a response of a remote local agent to the previous message. The remote local agents provide the amount of QoS that can be

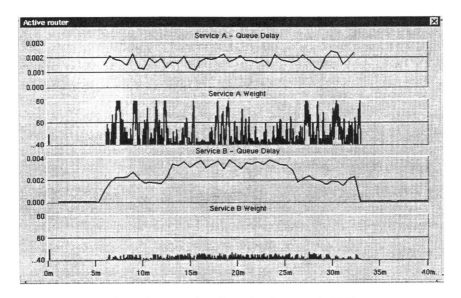

Figure 5. Agents-based weight adjustment in WFQ

provided by its nodes to recover from QoS shortage according to the local QoS reservations.

– **Allocate_QoS:** After processing all messages received from neighbor agents, the initial agent executes a QoS distribution algorithm to identify the amount of QoS to request from each neighbor node and then request this QoS using the Allocate_QoS primitive.

6. Simulating agents' behaviour

6.1. Typical scenario – video conferencing, video on demand

In order to validate and observe the behaviour of agents operating in an open networking environment, a computer model of an open network was developed. This was then used as a testbed (or prototyping environment) to observe the behaviour of the various real network agents during the course of their operations in undertaking a typical brokering scenario. The scenario we present consists of three competing networks, each identical in terms of the number of devices, topology and available resources. The individual networks consist of simulated switches interconnected in a small mesh topology.

Figure 6 shows an example where user wishes to download a movie from a remote video provider site. A SLA or GUI is provided to the user by the video service provider. The user sends an agent to the service provider for a movie list. The service provider asks the video provider for a movie list by sending an agent to it. The video provider processes the request and hands over the list to the service provider agent.

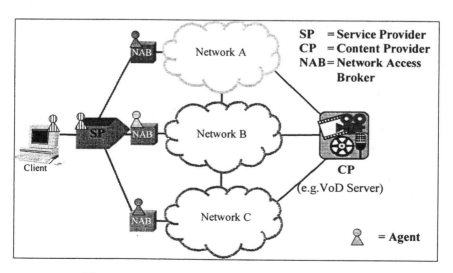

Figure 6. Network model for agent brokering environment

The service provider agent notifies the user that the list has been retrieved. To get a network connection the user sends a *service broker agent* to the service provider. The agent contains a SLA specifying the screen resolution, frame rate, duration and price issues. Upon the arrival of the request, the service provider manager asks the NABs (Network Access Brokers) of different networks (A, B, and C) for this particular service (Figure 7).

The service provider communicates with the NABs by *network broker agents* to get the quotations for the price according to the VPN routes plan. The network broker agent sends the information to the service provider about the QoS parameters (delay limit, jitter, max error/loss rate) to the service provider, to be passed onto the users. After obtaining the users' responses to the service offered, the active manager can then map the SLA requirements onto specific network resources. In order to support specific QoS targets through a VPN's route this will require creation of modifying resource flow specifications (in the case of RSVP), or provisioning DiffServ traffic class behaviours through the route. In our example, it may mean reconfiguring the scheduling schemes by sending active packets to the routers.

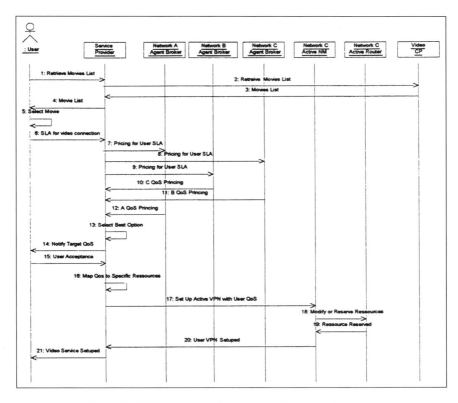

Figure 7. *UML sequence diagram of video connection setup*

The ObjectSpace™ Voyager ORB 3.0 [22] has been adopted for the implementation of our Agent prototype system. Voyager Agent API offers the ability to construct remote objects in the remote host and a set of control mechanisms that offer more flexible instructions on how the agent should terminate itself. There are two types of communication mechanisms, namely Method Calling and ObjectSpace™ by which the Voyager agents interact with each other. The former mechanism enables an agent to call methods of another agent provided the *calling* agent knows *a-priori* the method interface of the *called* agent. The latter mechanism enables the voyager agent to multicast an event message to other agents [22].

For the simulation we consider some universal customer behaviors. Firstly, the Customer will request the required bandwidth or QoS (Priority) and the price offered by each network provider. If the cheapest network provider cannot provide the require bandwidth, he/she will opt for the second cheapest, and so forth. During the simulation, end users' agents continually negotiate with network operators' agents for multimedia services such as voice, video and data. The users' service request rates were generated according to different Poisson distributions. Table 4 shows different classes of users' applications requesting services from the VPNs.

Table 4. Classes of users

Traffic Class	Request (per hour)	Mean BW Requested Unit per Connection	Session Time per Connection (mins)	Applications
1	70	2	3–10	VOIP/Internet Phone
2	15	30	10–60	MPEG2 Video/Video Conferencing
3	28	20	1–10	Internet Browsing

Figure 8 shows the accumulated bandwidth request profile of each class of users over time. Class 1 users' requests are short duration connections, but very frequent. Class 2 (video) users request high bandwidth connections over long periods of time. Class 3 users request services that provide bursty traffic profiles, such as web browsing.

Figure 9 shows the pricing history of the three acting network operators. Here, three network operators were trying to maximize the revenues by setting different BW prices at one time. At t>20mins, network A lowered its bandwidth price to 1 that caused a sharp increase in load over the measured link (shown in Figure 8). At t>50mins, network A increased its price dramatically and soon became much more expensive than, the others. As a result, a significant drop in traffic observed placed after t>75mins was most likely due to video subscribers leaving the network. At another time instant, t>110mins when network B's price remained constant, network A beat network B in price and diverted the traffic to its network. Notice that at time t>100mins, when network A was still the most costly network, traffics were coming into the network because the other two networks were saturated and were unable to provide the required bandwidth.

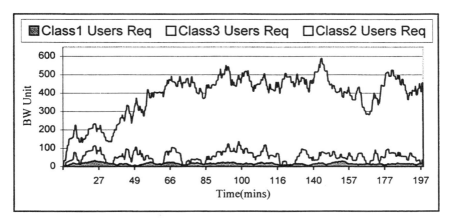

Figure 8. Multimedia users' bandwidth request profile – accumulated bandwidth

7. Conclusions and future work

In this paper we have presented an integrated framework for adaptive open networks based on contractual relationships between customers and a service provider represented by SLA. Future networks are expected to be open systems where end users or third party service providers can programme or customise network elements in order to obtain their required services. The framework developed considers a number of key technologies that may be integrated to allow dynamic modification of the services offered over a network. The technologies employed include active and programmable networks, agent technologies, MASIF, CORBA, and dynamically reconfigurable hardware. The framework also identifies different levels for the implementation of these technologies within the

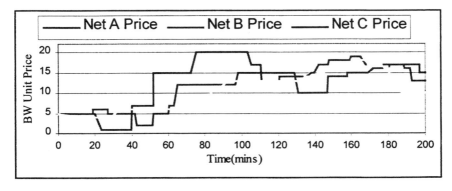

Figure 9. Bandwidth price bidding vs. time (minutes)

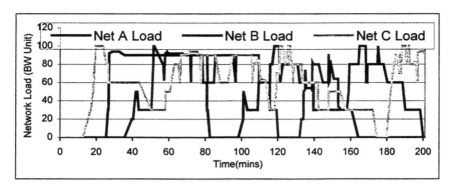

Figure 10. *Network load measured at one of the links (link2) vs. time*

network. An active network manager has been proposed at the network level to take care of network operations, policy management and security issues.

While the higher layers provide adaptation of the service requirements, the QoS policy (e.g., RSVP versus Diffserv), and even the choice of network operator, it is only at the network element level where these changes are realized. Dynamically reconfigurable hardware (for example, run-time reconfigurable FPGAs) is used to effect in real time the QoS target changes on each router in a VPN. Such reconfigurable network elements are termed 'active routers'. In operation, Delegated Agents acting on behalf of the service providers are dispatched onto a set of selected active routers by the Active Network Manager. Active packets from the Manager are used to reconfigure the scheduling schemes within the active routers as new SLAs or QoS updates occur. A new mechanism termed *Active Scheduling* has been developed to support dynamic reconfiguration of the schedulers. This mechanism allows delegated agents to perform a dynamic adjustment of session in active router to maintain the QoS required by end user applications. When the degradation is too important and not possible to resolve at the local agents schema, delegated agents located in the same or different active routerscan cooperate in order to find out a distributed solution by reallocating QoS between routers.

It is expected that future network infrastructure will conform to a layered market model whereby sufficient alternatives exist that allow services to be traded as commodities. The paper also describes how agents can facilitate brokering activities for network resources and services in such an open heterogeneous network environment. Simulations have been developed to model the behavior of the various agents operating in such open networks, and these have shown that the use of agents can introduce a much greater dynamic into the provision of network services. The framework described provides a platform for future research into the application and validation of agent technologies for next generation network management and control.

The work following is to extend the architecture to make the framework more programmable at the end user application boundary. Thus, it will permit applications to express in more detail the required communication service they need and also the exact behavior they expect from the network according to parameters such as price, QoS, security, etc. The idea is to define different profiles of communication services that can be easily proposed by a service provider and adapted to future need and business objectives. Service providers can collaborate together to provide an end-to-end service. This will require automatic interoperability between service provider platforms to support the combination of services to be provided to the end user.

Acknowledgements

The authors gratefully acknowledge support and financial assistance provided by Nortel Networks under the JIGSAW project.

REFERENCES

[1] BISWAS J. et al., "IEEE P1520 Standards Initiative for Programmable Network Interfaces", *IEEE Communications Magazine*, Vol. 36, N° 10, October 1998, p. 64–70.

[2] BREUGST M. and MAGEDANZ T., "Mobile Agents-Enabling Technology for Active Intelligent Network Implementation", *IEEE Network Magazine*, Special Issue on Active and Programmable Networks, May/June 1998, Vol. 12, N° 3.

[3] TENNENHOUSE D.L. et al., "A Survey of Active Network Research," *IEEE Communications Magazine*, January 1997, p. 80–86.

[4] CAMPBELL A.T. et al., "A Survey of Programmable Networks," *ACM SIG-COMM Comp. Commun*, April 1999.

[5] BHATTACHARJEE S. et al., "Directions in Active Networks", *IEEE Communications Magazine*, October 1998, p. 72–78.

[6] TENNENHOUSE D.L. and WETHERALL D., "Towards an Active Architecture", *Proceedings of Multimedia Computing and Networking (MMCN)*, San Jose, January 1996.

[7] SMITH J.M. et al., "Activating Networks: A Progress Report", *Computer*, April 1999, p. 32–41.

[8] WETTHERALL D. et al., "Introducing New Internet Services: Why and How", *IEEE Network*, July 1998.

[9] FAN Z. and MEHAOUA A., "Active Networking: A New Paradigm for Next Generation Networks?", *Second IFIP/IEEE Int. Conference on Management of Multimedia Networks and Services (MMNS'98)*, France, November 1998, p. 1–9.

[10] LESLE et al., "The Tempest – A Practical Framework for Network Programmability", *IEEE Network Magazine*, May/June 1998, p. 20–28.

[11] TENNENHOUSE D.L et al., "ANTS: A Toolkit for Building and Dynamically Deploying Networking Protocols", *IEEE OPEN ARCH '98*, San Francisco, CA, April 1998, p. 1–12.

[12] MARSHALL A. and SEZER S., "The Influence of Cumulative Switch Delay in Multiple Service Class Networks", *International Journal of Computers and Applications*, 2000, Special Edition.

[13] SEZER S., MARSHALL A., WOODS R.F. and GARCIA-PALACIOS F., "Buffer Architectures for Predictable Quality of Service at the ATM Layer", *IEEE Global Telecommunications Conference, GLOBECOM '98*, Sydney, Australia, November 1998.

[14] GARCIA-PALACIOS F., MARSHALL A., SEZER S. and CHIENG D., "QoS Analysis of a Wireless ATM Network Access Point", *ICC '99*, Vancouver, Canada, June 1999.

[15] HUSSAIN S.A. and MARSHALL A., "An Active Scheduling Policy for Programmable Routers", *Sixteenth UK Teletraffic Symposium – Management of QoS – The New Challenge*, Harlow, May 2000.

[16] MARSHALL A., "Dynamic Network Adaptation Techniques for Improving Service Quality", *Networking 2000*, Paris, May 2000.

[17] CHIENG D. *et al.*, "A Mobile Agent Brokering Environment for The Future Open Network Marketplace", *Seventh International Conference on Intelligence in Services and Networks (IS&N2000)*, Athens, Greece, Febuary 2000.

[18] AGOULMINE N., DRAGAN D., GRINGEL D., HALL J., ROSA E., TSCHICHHOLZ M., "Trouble Management for Multimedia Services in Multi-Provider Environments", *International Journal on Network and Service Management*, John Wiley & Sons, Vol. 8, January 2000, p. 99–123.

[19] BRADEN R., ZHANG L., BERSON S., HERZOG S. and JAMIN S., "Resource ReSerVation Protocol (RSVP), Version 1 Functional Specification", *RFC 2205*, Network Working Group, IETF, September 1997 (www.ietf.org).

[20] BLAKE S., BLACK D., CARLSON M., DAVES E., WANG Z. and WEISS W., "An Architecture for Differentiated Services", *RFC 2475*, Network Working Group, IETF, December 1998 (www.ieft.org).

[21] White Paper – "QoS Protocols & Architecture", Stardust.com, Inc., July 8, 1999 (www.qosforum.com).

[22] http://www.objectspace.com/products/voyager/

[23] Recommendation V.90 – 56,000 bits per second duplex modem standardised for use in the general switched telephone network, ITU-T, Geneva 1999.

[24] Xilinx VIRTEX VCX100 Field Programmable Gate Arrays, Advance Product Specification, May 1999 (Version 1.50, Xilinx Inc.).

[25] VILLASENOR J. and MANGIONE-SMITH W.H., "Configurable Computing", *Scientific America*, June 1997, p. 54–59.

[26] OMG Mobile Agent System Interoperability Facility (MASIF) Specification, ftp://ftp.omg.org/pub/docs/orbos/97–10–05.pdf

[27] FERRARI D. and VERMA D., "A Scheme for Real-Time Channel Establishment in Wide-Area Networks", *IEEE Journal on Selected Areas inn Communications*, vol. 8, N° 3, April 1990, p. 368–379.

[28] NAIT-ABDESSELAM F., AGOULMINE N., KASIOLAS A., "Agent-Based Approach for QoS Adaptation in Distributed Multimedia Applications over ATM", in *Proceedings of International Conference on ATM*, Colmar, France, June 1998.

[29] FERRARI D. and VERMA D., "A Scheme for Real-Time Channel Establishment in Wide-Area Networks", *IEEE Journal on Selected Areas in Communications*, vol. 8, N° 3, April 1990, p. 368–379.

[30] AURRECOECHEA C., CAMPBELL A. *et al.*, "A Review of Quality of Service Architectures", *ACM Multimedia Systems Journal*, November 1995.

Chapter 6

Adaptive multicast group management for distributed event correlation

Ehab Al-Shaer

Multimedia Networking Research Laboratory, School of Computer Science,
Telecommunications and Information Systems, DePaul University, Chicago, USA

1. Introduction

In distributed management environments, large numbers of events (notifications or alarms) are generated by network or system components during their execution and interaction with external objects (e.g. users or processes). These events must be monitored, classified, filtered and analyzed to accurately determine the actual cause of the problems such as faults, security threats or performance bottlenecks. This process is called event correlation and it is significantly important for fault, performance and security management. The manner in which events are generated is complex and represents a number of challenges for real-time event correlation. Correlated events are generated concurrently and can occur at multiple locations distributed throughout the environment. Furthermore, the large number of managed entities, the geographical distribution, and the dynamic behavior inherent with the next-generation distributed services increase the difficulty of addressing critical issues in distributed event correlation, such as scalability, performance bottlenecks, and application perturbation.

Next-generation event correlation services must be scalable and dynamic to handle large numbers of managed objects efficiently. In distributed event correlation systems, a group of monitoring agents are used to exchange event notifications and perform correlation collaboratively. Hence, an agent may need to forward detected event notifications to a group of agents for further analysis or to a group of managers interested in this event. Similarly, managers also communicate with a group of agents in order to distribute the event correlation tasks. Using group multipoint communication is evidently significant in this environment to improve the scalability and performance of the distributed event correlation systems. However, because agents' group membership must be dynamically changed based on the event notification and correlation process, a highly configurable and dynamic group management is required in order to provide efficient group communications.

This paper presents a dynamic group management framework based on IP multicast to support scalable distributed event correlation. The proposed framework uses the event correlation information to re-configure the multicast group formation dynamically. While this minimizes the number of messages and processing delay compared with point-to-point communication, this group management also allows for the optimal formation of multicast groups in distributed monitoring applications. The presented group management and communication framework was implemented in the HiFi monitoring system that employs a hierarchical event correlation [ALS 00, ALS 90].

Although several distributed event correlation techniques were proposed (e.g., [ALE 96, GAR 96, GAR 98, JOR 89, WU 98]), exploiting group communication to improve the scalability and performance of event correlation is not sufficiently addressed. Also, a number of studies propose integrating group communication in distributed management [AMI 92, LEE 95, PAR 99, SCH 96]. However, as discussed in Section 6, they do not address the issues of dynamic group management which thereby limits the efficiency of these systems in distributed event correlation environment.

This paper is organized as follows: Section 2 gives an overview of a HiFi monitoring system which includes the model, language and the architecture; Section 3 presents our dynamic group management and communication framework for distributed event correlation systems, Section 4 presents the agents synchronization and state consistency protocol; Section 5 explains the automatic bootstrap mechanism for establishing the agents hierarchy; Section 6 discusses related work; Section 7 presents the summary and concluding remarks.

2. HiFi monitoring system overview

HiFi employs a hierarchy of collaborative agents that receive events as alarms or as notifications from instrumented programs. The monitoring agents filter and analyze the generated event based on correlation rules defined by the managers in filter scripts. In this section, we give an overview of the HiFi monitoring architecture and refer to [ALS 90, ALS 99] for more information.

Monitoring model and language: HiFi is an event-demand-driven monitoring model. In other words, the producer behavior is observed based on the event generated (event-based) and on the monitoring requests (subscription-based). Therefore, events received in the monitoring system are classified based on exiting monitoring requests (called filters). The monitored programs (called event producers) continuously emit events that express the execution status. An event is a significant occurrence in a system that is represented by a notification message which typically contains information that captures event characteristics such as event type, event generation time, event source. There are two types of events used in our model: primitive events which are based on a single notification message, and composite events which depend on more than one notification message. The event specification language in HiFi represents the event format (notification) as a variable sequence of event attributes determined by the user but it has a fixed header used in the monitoring process [ALS 00].

In HiFi, managers (called event consumers) specify their monitoring demands by sending a filter program dynamically via the subscription process which configures the monitoring agents accordingly. A filter is a set of predicates where each predicate is defined as a boolean-valued expression that returns true or false. Predicates may be joined by logical operators (such as AND and OR) to form an expression. A typical HiFi filter consists of three major components: the event expression which specifies the relation between the interesting events, filter expression which specifies the event attributes value or the relation between the attributes of different events, and the action to be performed if both event and filters expressions are true. If an event is detected, the action specified in the filter such as forwarding the monitoring information to the corresponding consumers is performed. For example, assume a manager requests detecting warning events, *AudioWarning* and *VidWarning*, caused by Audio and Video processes in a distributed system respectively only if they are generated by the same machine. This is represented in the following filter:

FILTER= [(*AudioWarning* ∧ *VidWarning*)];
[(*AudioWarning.Machine*=*VidWarning.Machine*)];
[FORWARD];Warnings_Correlation_Filter.

HiFi also provides the Environment Specification Language (ESL) which the managers use to describe the application distribution in the network such as which process exists in which machine and domain. The ESL is used for automating the establishment of the agent hierarchy as we describe later in Section 5. The formal specification of HiFi monitoring language and examples are presented in [ALS 99].

Hierarchical event correlation: HiFi employs a hierarchical event filtering-based monitoring architecture to distribute the monitoring load in application environment. The task of detecting primitive and composite events is distributed among dedicated monitoring programs called monitoring agents (MA). MA is an application-level monitoring program that runs independently of other applications and communicates with the outside world (including producers and consumers) via message-passing. HiFi has two types of MAs: local monitoring agents (LMA), and domain monitoring agents (DMA). The former is responsible for detecting primitive events generated by local applications in the same machine while the latter is responsible for detecting composite events which are beyond the LMA scope of knowledge. One or more event producers (i.e., processes) are connected to a local LMA in the same machine. Every group of LMAs related to one domain (geographical or logical domain) is attached to one DMA. These DMAs are also connected to higher DMAs to form a hierarchical structure for exchanging the monitoring information. Figure 1 shows the hierarchical agent architecture in HiFi.

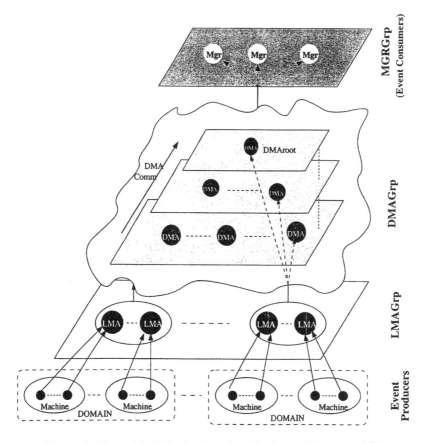

Figure 1. *Hierarchical filtering-based monitoring architecture (HiFi)*

Subscription process: Based on a user's monitoring request, the monitoring system determines the appropriate agent or set of agents within the hierarchy to be tasked with inspection and evaluation of application events. The monitoring system uses fine grain decomposition and allocation mechanisms to ensure that filtering tasks are efficiently distributed among the monitoring agents and minimize events propagation in the network. The monitoring process starts by a consumer sending a filter program that describes the monitoring request to the local MA. The filter is validated and decomposed into subfilters through the decomposition process in such a manner that each one represents a primitive event [ALS 90]. The filter expression is also decomposed into subexpressions where each one is contained within a domain. Then, each decomposed subfilter or subexpression is assigned to one or more LMAs or DMAs through the allocation process based on the event sources and application distribution. The decomposition and allocation process are described in [ALS 90]. The monitoring system also determines the optimal DMAs for evaluating the event and filter subexpressions of a filter program.

This architecture alleviates any performance bottlenecks or scalability problems by distributing the monitoring load among MAs and limiting the events' propagation to the originating sources [ALS 90].

Event generation and code instrumentation: Events are generated from the running program after instrumenting the program code with sensors or probes. In order to facilitate the instrumentation process, HiFi provides the Event Reporting Stub (ERS) routine which is a library linked with the program during compilation. During the program execution, ERS constructs and generates the triggered events as they occurred and sends them to its LMA. Each ERS connects to LMA with a communication channel via UNIX sockets [STE 96]. The upstream communication is used to send primitive events from ERS to an LMA, and the downstream is used to receive control information such as activation and deactivation of event reporting [ALS 99].

3. Group communication for distributed event correlation

As described in Section 2, HiFi employs a group of autonomous and collaborative agents (LMAs and DMAs) that communicate with each other and with the managers in order to classify and correlate events. We use IP multicast to provide the group communications in a HiFi distributed event correlation service. IP multicast is considered the de facto standard of multi-point group communication in the Internet. IP multicasting enables the sending of packets to a group of receivers without duplicating them at the source. Multicast packets are, instead, duplicated automatically by multicast routers that are close to the receivers in the network. Multicast receivers can start and stop receiving multicast packets dynamically through join and leave multicast operations respectively. The monitoring agents and the managers use the Reliable Multicasting Server (RMS) described in [ALS 97] for this purpose. This section describes the group management framework and communication used in the HiFi event correlation service.

3.1 LMA-DMA group communication

This communication is used to forward detected primitive events from LMAs to DMAs. After the decomposition and allocation process, each LMA knows which primitive event to forward and each DMA knows which primitive event to request in order to evaluate the delegated subexpression. As a result, every DMA uses the names of the requested events to form/join multicast groups such that the event name is the prefix and "Grp" is the suffix. For example, if a DMA requires *Event X* notification in order to evaluate delegated subexpression, this DMA joins the multicast group called *EventXGrp*. Similarly, LMAs use the names of detected events as the group names to which event notifications are multicasted. For example, if *EventX* event is detected by an LMA, this LMA multicasts this event to the *EventXGrp* group. Consequently, this causes the *EventX* event to be forwarded to all DMAs interested in receiving this event. The event and filter information such as event names and filter subexpressions are distributed by the manager after the decomposition and allocation processes and during the agent hierarchy boot-strap as described in Section 5. Notice that, in IP multicast, an LMA does not need to be a member in any multicast group in order to send multicast events to the DMA groups (called open groups). In RMS, non-members are required to use 'connect' (e.g., connect GrpName) in order to send messages to multicast groups reliably. For this reason, LMAs use the RMS connect request to establish reliable multicast connections with multicast groups designated as ($<$ *primEventName* $>$ *Grp*) as described above in order to forward primitive events.

Moreover, a DMA may occasionally need to multicast control or administrative information to other LMAs in the same domain for load adaptation or accommodating new application entities [ALS 99]. In order to facilitate this domain-based communication, LMAs within the same domain (i.e., sharing the same DMA) join a multicast group designated by the domain name as a prefix and "Grp" as a suffix. The domain information such as domain names and managed objects distribution is specified by the manager using the Environment Specifications Language (ESL).

3.2 DMA-DMA group communication

As a result of decomposition and allocation process described in Section 2, the event correlation expression represented in the filter expression (FX) is fragmented and distributed among a number of DMAs in the hierarchy [ALS 90]. This results in minimizing event propagation and distributing the monitoring load among the agents. Hence, if a DMA evaluates its delegated subexpression, it then forwards the evaluation result to one or more higher DMA in the hierarchy which, in turn, uses this result to continue the evaluation of the filter expression. For this reason, DMAs join multicast groups that correspond to the delegated subexpressions. The names of the multicast groups are directly derived from the subexpression itself such that the expression is used as a prefix after replacing AND and OR operators by "A" and "O" respectively, and "_Grp" is used as a suffix. For example, if the subexpression $E_1{\wedge}E_2$ is delegated to DMA_x and the evaluation result must be forwarded to DMA_y, then DMA_x joins E_1Grp and E_2Grp multicast groups (as described before) to receive information about these events, and DMA_y joins the multicast group $E_1AE_2\_Grp$ to receive information about the subexpression evaluation results. In addition, DMA_x sends the evaluation result of this subexpression to $E_1AE_2\_Grp$ group. Similarly, the multicast communication is used for forwarding composite events (i.e., subexpression evaluations) to higher DMAs in the hierarchy until the process of event correlation completes.

3.3 Managers-agents group communication

HiFi monitoring systems allow a group of managers to share the results of monitoring operations in a scalable manner. The monitoring agents forward event notifications to one or more managers according to their subscription requests. On the other hand, a manager may communicate with a group of agents in order to distribute monitoring demands. Therefore, a many-to-many communication model is required between managers and agents. In our group management framework, every manager joins Mgr-Grp multicast group. LMA and DMA agents use this group to send control or administration information to managers. Moreover, each manager joins multicast groups based on its submitted filters such that the filter name is used as a prefix and "Grp" as a suffix in the group name (<FilterName>Grp). More than one manager can share the same filter (e.g., MyFilter) by simply joining the same multicast group associated with this filter name (*MyFilterGrp*). When the requested event correlation of a filter is detected, the monitoring agents send a notification to <FilterName> Grp group of this filter (e.g., *MyFilterGrp*) which thereby is received by managers joining this group simultaneously. However, this may cause two or more managers submitting different filter correlations to join the same multicast group because they coincidentally select the used filter name. To avoid this multicast group conflict between managers, the name of a submitted filter must be unique in the monitoring environment. More specifically, if a manager submits a filter successfully (called the filter owner), no other manager is allowed to submit a filter with the same name as described later in this section. For security reasons, no manager except the filter owner is permitted to modify or delete a filter. The filter owner can also specify if the submitted filter can be sharable by other managers or not as described later in this section.

One way to provide a collision-free multicast group allocation is to use distributed domain proxies as described in [PEJ 95]. However, this technique is inefficient in our case because managers are usually located at different domains, thereby a proxy is required for each manager. This is in addition to the complexity inherent with maintaining the proxies in a distributed environment. We, instead, developed a simple and fully distributed protocol to allocate filter names (or multicast groups) exclusively by the manager. The algorithm of this protocol is outlined in Figure 2. Before a manager submits a filter, the manager first searches for the filter name in its local monitoring knowledge-base (Check–MKB()). If it is not found, then the manager multicasts a filter name and specification such as filter correlation and sharing status to the *MgrGrp* group. When managers receive this information, they immediately add it to their local knowledge-base along with the IP address of the sending manager. Therefore, managers are always aware that the active filters exist in the monitoring environment. After multicasting the filter

Input: A filter specification (FX, and a filter name)
Output: 1 or 0 for successful and unsuccessful submissions

SafeSubmit(FilterInfo)
      if (Check_MKB(FilterInfo.Fname)== NOTFOUND)
          SendMcastToMgrGrp(FilterInfo);
          Wait($RTT * REXMT_{max} + \alpha$); /* Receive while waiting */
          if (Check_MKB(FilterInfo.Fname) == NOTFOUND)
               Submit(FilterInfo);
               return 1;
          else
               Wait($RTT * REXMT_{max} + \beta$); /* Receive while waiting */
               if (Fname_was_Submitted)
                    return 0;
               else
                    HighestIP = GetHighestIP();
                    if (HighestIP <= ThisIP)
                        Submit(FilterInfo);
                        return 1;
                    else
                      return 0;
      else
          return 0;

Figure 2. *Safe filter submission algorithm*

information, the manager waits for more than the maximum allowable re-transmission timeout period ($RTT_{max} * REXMT_{max} + \alpha$ where α is the maximum database update/check time). During the waiting period, managers receive and process multicast messages. When the timer expires, managers re-check the local database again in order to ensure that no other manager is attempting to use the same filter name recently. If the second check passes successfully, then the manager submits the selected filter name which results in multicasting the filter information to *MgrGrp* group. Otherwise, it means another manager attempts to submit a filter with the same name simultaneously. In this case, the manager waits again until it receives the submitted filter from *MgrGrp* for a timeout period ($RTT_{max} * REXMT_{max} + \beta$ such that $\beta = \alpha$ + submission time) or the manager that has the highest IP number, HighestIP, (and port number if they are from the same machine) is granted this filter name. In the latter case, managers update their MKB at the same time and they all find a match in the local KMB. The proof of correctness of this algorithm is presented in Appendix A.

On the other hand, when a manager deactivates a filter (e.g., MyFilter), it then leaves *MyFilterGrp* and sends a multicast message to *MgrGrp* to indicate this deactivation which causes managers to delete this filter from its local knowledge-base.

Finally, to enable managers-to-agents group communication, both LMAs and DMAs use RMS to join *LMAGrp* and *DMAGrp* multicast groups respectively. Managers forward the monitoring information such as events, filters and application environment to the LMAs and DMAs via multicasting it to *LMAGrp* and *DMAGrp* respectively. However, when a manager multicasts the filter decomposition information to *LMAGrp* and *DMAGrp* groups, it also indicates in the message the delegated tasks and the associated agents. When an agent receives this message, it looks up its own delegated tasks, if any, and acts accordingly.

Although the proposed framework uses reliable multicasting, unreliable multicast can also be used to provide the same service. Using a reliable multicasting simplifies the monitoring process and provides an accurate and deterministic view (no event loss) to the event correlation engine. This is particularly important in distribued systems managements. (See Figure 3.)

4. Agents' state consistency protocol

After the monitoring agents' hierarchy is established (as described in Section 5), manager(s) can start the subscription process via adding, modifying or deleting filters on-the-fly. Adding, modifying or deleting filter components are performed on different agents simultaneously using group communication. This may cause a state inconsistency in the monitoring agents' environment due to communication delay or failures.

To solve this problem, agents' state update must be atomic (i.e., an agent failure causes all other agents to abandon the process) and synchronized (i.e., agent waits until all participating agents commit the state update). We developed agents state consistency protocol as part of the subscription process to provide an atomic and synchronized subscription operation. The protocol is based on reliable multicasting and group management described in Section 3 and its state diagram is depicted in Figure 4. The subscription component in the manager program parses, decomposes filter programs, and multicasts the delegated tasks (subfilters) to the

Group Name	Member	Sender	Purpose
MgrGrp	manager	manager	group synchronization and admin info
	manager	*DMA_root*	agents' hierarchy confirmation
DMAGrp	DMA	manager	forward event,filter and environment info
LMAGrp	LMA	manager	forward event,filter and environment info
< *EventName* > *Grp*	DMA	LMA	forward the primitive events notification
< *FilterName* > *Grp*	manager	DMA or LMA	forward the event correlation info
< *DomainName* > *Grp*	LMA	DMA	forward control info
< *SubExpr* > _*Grp*	DMA	DMA	forward FX subexpression evaluation

Figure 3. Monitoring agents' multicast groups

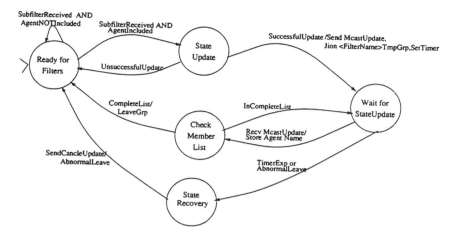

Figure 4. *Agents' state consistency protocol state diagram*

monitoring agents. The delegated subfilter messages contain the filter name, the decomposed subfilters, and a list of agents IDs needed for this monitoring task. When an agent receives a Subfilter message that contains its ID (*MachineName.DomainName*), it performs the filter composition to insert the delegated subfilters into its internal filtering representation (StateUpdate state). If the agent updates its state successfully, it then uses the filter name included in the message to join < *FilterName* > *TmpGrp* multicast group and sends a join notification, McastUpdate, to this group. When other agents in the group receive the McastUpdate message, they add the agent's ID to their local repositories (MemberList). Similarly, agents continue adding IDs into their local repositories until they contain all agents found in the subfilter message. This implies that all agents found in the subfilter message have completed their state update successful and thereby they can leave < *Filter Name* > *TmpGrp* multicast group and activate the delegated filters.

On the other hand, if an agent fails to update its state or join a group, one agent at least will time out (i.e., timer expires) and send a multicast message to the < *FilterName* > *TmpGrp* group to cancel and recover the state update. Every agent sets up a timeout timer right after joining < *FilterName* > *TmpGrp* group for a time period specified as:

$TimeOut = (2 * REXMT_{max} * RTT_{max}) + \alpha$ such that RTT_{max} is the maximum round trip delay in the network, $REXMT_{max}$ is the maximum number of retransmissions in RMS. Therefore, $2 * REXMT_{max} * RTT_{max}$ is the maximum aggregate retransmissions time in the network for both the Subfilter message and the McastUpdate. And α is the maximum state update processing time. Notice that the *TimeOut* expression represents the worst case so that agents never time out before the completion of the protocol operations successfully. In case of agents' crashes, RMS protocol detects this event and notifies all members in the group instantaneously which causes the agents to revoke the state update operation.

At the end of this operation, managers get notified about the results of their subscription, (i.e., confirmed or aborted) by a monitoring agent which sends the result the to < *FilterName* > *Grp* group. One of the main advantages of this protocol is simplicity and minimal overhead compared to other distributed algorithms such as two-phase commit protocol [TAN 93].

5. Automatic agents' bootstrap mechanism

To avoid the complexity of constructing and administrating the agents' hierarchy and groups, we developed a protocol based on the group management framework described before that automates creating, allocating and setting up the agents' hierarchy dynamically without the involvement of the users. This service significantly facilitates the administration of HiFi management systems. Figure 5 shows the protocol interaction diagram between the parties of this protocol operation. In this section, we give a brief description for this agents'

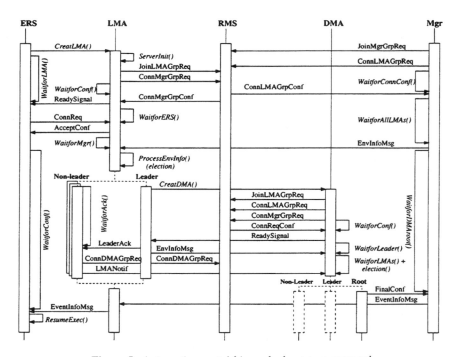

Figure 5. *Automatic agents' hierarchy bootstrap protocol*

hierarchy bootstrap protocol. Although in the case study below we use HiFi for monitoring distributed systems, the same technique is applied in any other event-based or trap-based management environment. After the application is instrumented, the agents' hierarchy is established and the consumers send their subscriptions.

1. (Manager Program starts.) When a manager program starts it joins MgrGrp and connects to LMAGrp (as described before in Section 3, 'connect' is used to send to a multicast group reliably but 'join' is used for receiving and sending to multicast groups). Then, the manager waits for all LMAs to start and connect to MgrGrp. RMS sends a notification to the manager (MgrGrp) whenever a member joins or connects to a group.

2. (Instrumented Program starts.) When the instrumented program starts, the ERS creates (fork ()[1]) an LMA and waits for Ready signal from this LMA.

3. (LMA starts and an LMA-ERS connection established.) After the LMA starts it joins LMAGrp and sends a SIGUSR UNIX signal to ERS which consequently establishes a UNIX socket connection [STE 96] with the LMA. When all LMAs are created and ready, the manager multicasts the environment information (EnvInfo) to LMAGrp. Notice that the managers know about the total number of LMAs from the environment specifications. Late-join LMAs can also get connected as described in [ALS 99].

(a) (LMA election process starts.) Upon receiving EnvInfo from a manager, LMAs go through an election process based on the position of LMA name/ID in the EnvInfo table. In particular, the first LMA name in the LMA's list of each domain is the LMA leader. As a result, LMAs are divided into two groups: a leader group that contains the LMA leaders for all domain, and a non-leader group that contains the other LMAs in each domain.

(b) Each LMA leader creates the designated DMA for this domain, forwards the environment information to it, and sends an acknowledgment to non-leader LMAs to announce DMA creation. Upon receiving this acknowledgment, non-leader LMAs connect to the DMAGrp group.

4. (DMA starts and the election process.) After all LMAs in the domain are connected to the DMAGrp, the DMAs go through the same election process used by the LMA which classifies DMAs into: DMA leader, DMA non-leader and DMA root. The first two groups (DMA leader and non-leader) follow the same steps described for LMA leader and non-leader. This implies that every DMA leader creates its containing or higher DMA (called superDMA) and this hierarchical construction continues until DMA root is created. When the DMA root starts, it immediately sends a final confirmation to the manager (MgrGrp) confirming the completion of the agents' hierarchy.

5. The Manager then multicasts the event information (EventInfoMsg) to the LMAGrp and ERS. ERS uses the events information to construct and send events

---

[1] We assume that the agents binary exists in the same location or file server where the monitored programs exist.

notifications and LMAs use the event information to join primitive event multicast groups as described in Section 3.

6. ERS resumes the program execution and the event reporting process. The LMAs and DMAs are completely set up in their groups and ready for filter delegations.

The protocol scales well with the number of agents since DMAs at the same level and LMAs operate concurrently and the effect of the hierarchy height is minimal. It is important to mention that process crashes or abnormal levels from the multicast groups (ERS, LMAGrp, DMAGrp and MgrGrp) are immediately propagated to the rest of the agents and cause the agents to abandon this process and quit. This guarantees that the final confirmation is sent only if the entire agent hierarchy is constructed successfully.

6. Related work

Our related work study focuses on the systems that attempt to develop or exploit group communication for distributed network management applications.

A framework for using IP multicast group communication with SNMP is proposed in [SCH 96]. This framework provides a primitive group membership structure. It also uses an election algorithm to choose a master agent that facilitates the group communication between other SNMP agents. The SNMP trap messages over IP multicast are utilized for exchanging control information. This framework is an example of developing autonomous SNMP agents using IP multicasting. However, it does not present a general framework for integrating IP multicasting in standard SNMP agents because (1) it lacks the flexibility of re-configuring the multicast group structure and the communication model dynamically based on the application needs, (2) it requires major changes in the SNMP agent in order to use this framework, and (3) due to the periodic trap messages and the election process, this approach may cause a considerable overhead on the SNMP agent.

In [LEE 95], a reliable group communication protocol for distributed management that preserves message ordering and atomicity was described. This protocol uses a hierarchy of servers and logical timestamps to ensure reliability and causal ordering of group delivery. It seems that this architecture was developed over unicast connections, instead of employing IP multicasting, to emulate group communication. This significantly limits the performance and scalability of the proposed architecture. Furthermore, this work does not address the issue of group management of monitoring agents which is the main theme of this paper.

Another group communication infrastructure based on Transis [AMI 92] group communication system was proposed in [AMI 96]. It supports an efficient solution for some distributed system management applications such as software installation, simultaneous remote execution and configuration management on a cluster of servers. A monitor program uses Transis to communicate with a group of management servers that perform system management tasks reliably and consistently. A similar framework is proposed in [PAR 99] to use IP multicasting for control and management of distributed applications. Although these systems show the advantage of using group communication for some targeted management applications, they do not provide a general group management and communication framework for distributed event filtering applications. In addition, using a distributed system toolkit such as Transis as a core element limits the usability of the system particularly for Internet-based network management.

In conclusion we found in the course of our investigation that employing an efficient group management and communication for distributed event correlation was not sufficiently addressed by previous related work.

7. Conclusion and future work

Employing efficient group management and communication is significantly important for supporting distributed event correlation services. This paper presents a new group management framework based on the IP multicast standard for distributed event correlation that exhibits the following key advantages:

– It supports a dynamic and scalable monitoring information dissemination to managers and agents efficiently compared with unicast communication.

– It provides a fine-grain group communication that enables agents to disseminate monitoring information based on the event correlation tasks and delegations. This provides for an optimal multicast delivery among agents and managers.

– The presented multicast management framework facilitates distributing the correlation tasks among a group of agents which minimizes the monitoring latency and eliminates performance bottlenecks in the event correlation process.

– The presented framework also supports agents' synchronization protocol that ensures agents' state consistency during group communication, a distributed algorithm that enables managers to share monitoring views (results) without multicast group conflict, and a bootstrap mechanism that facilitates the agents' creation and administration.

– IP multicasting improves the robustness and survivability of the system significantly in the presence of agents or network failures. In the case of network partitioning or agents malfunctioning, other agents can communicate and negotiate recovery procedures. This is unlike point-to-point TCP connections where a large number of backup connections is required.

One limitation of the described approach is the potential of creating too many groups that may cause high resource consumption (i.e., socket descriptors). However, this is not generally anticipated unless thousands of different primitive events (or alarms) are defined in each managed object which is very unlikely the case in typical enterprise network or system management systems. Examples of future issues to be explored include supporting soft real-time monitoring, integrating event ordering, using customized reliability and providing group fault recovery.

REFERENCES

[ALE 96] ALEXANDER S., KLIGER S., MOZES E., YEMINI Y. and OHSIE D., "High Speed and Robust Event Correlation". *IEEE Communication Magazine*, pp. 433–450, May 1996.

[ALS 90] AL-SHAER E., ABDEL-WAHAB H. and MALY K., "HiFi: A New Monitoring Architecture for Distributed System Management". In *Proceedings of International Conference on Distributed Computing Systems (ICDCS'99)*, pp. 171–178, Austin, TX, May 1990.

[ALS 97] AL-SHAER E., ABDEL-WAHAB H. and MALY K., "Application-Layer Group Communication Server for Extending Reliable Multicast Protocols Services". In *IEEE Int. Conference on Network Protocols*, pp. 267–274. Atlanta, GA, October 1997.

[ALS 99] AL-SHAER E., ABDEL-WAHAB H. and MALY K., "Dynamic Monitoring Approach for Multi-point Multimedia Systems". *International Journal of Networking and Information Systems*, pp. 75–88, June 1999.

[ALS 00] AL-SHAER E., "Active Management Framework for Distributed Multimedia Systems". *Journal of Network and Systems Management (JNSM)*, March 2000.

[AMI 92] AMIR Y., DOLEV D., KRAMER S. and MALKI D., "Transis: A Communication Subsystem for High Availability". In *IEEE Workshop on Fault-Tolerant Parallel and Distributed Systems*, pp. 76–84, 1992.

[AMI 96] AMIR E., BREITGAND D., CHOCKLER G. and DOLEV D., "Group Communication as an Infrastructure for Distributed System Management". In *Proceedings of Third International Workshop on Services in Distributed and Networked Environments*, pp. 84–91, July 1996.

[GAR 96] GARDNER R. and HARLE D., "Methods and Systems for Alarm Correlation". In *Global Telecommunications Conference (GLOBECOM '96)*, vol. 1, pp. 136–140, May 1996.

[GAR 98] GARDNER R. and HARLE D., "Pattern Discovery and Specification Techniques for Alarm Correlation". In *Network Operations and Management Symposium (NOMS'98)*, vol. 3, pp. 713–722, March 1998.

[JOR 89] JORDANN J.F. and PATEROK M.E., "Event Correlation in Heterogeneous Networks Using the OSI Management Framework". In *Integrated Network Management I*, pp. 365–379, IFIP, 1989.

[LEE 95] LEE K.-H., LEE J.-K. and KIM H.-S., "A Multicast Protocol for Network Management System". In *Proceedings of IEEE International Conference on Information Engineering*, pp. 364–368, June 1995.

[PAR 99] PARNES P. AND SCHEFSTORM D., "Real-Time Control and Management of Distributed Application using IP-Multicast". In *Proceedings of the IEEE/IFIP International Symposium on Integrated Network Management (IM)*, pp. 901–914, 1999.

[PEJ 95] PEJHAN S., ELEFTHERIADIS A. and ANASTASSIOU D., "Distributed Multicast Address Management in the Global Internet". *IEEE Journal on Selected Areas in Communications*, pp. 1445–1456, October 1995.

[SCH 96] SCHONWALDER J., "Using Multicast-SNMP to Coordinate Distributed Management". In *Proceedings of Second IEEE International Workshop on Systems Management*, pp. 136–141, March 1996.

[STE 96] STEVENS W. R., "*TCP/IP Illustrated, Volume 3: TCP for Transactions, HTTP, NNTP and the UNIX Domain Protocols*". Addison-Wesley, Reading, Massachusetts, 1996.

[TAN 93] TANENBAUM A. S., "*Modern Operating Systems*". Prentice Hall, Englewood Cliffs, NJ, 1993.

[WU 98] WU P., BHATNAGAR R., EPSHTEIN L. and SHI M. B. Z., "Alarm Correlation Engine (ACE)". In *Network Operations and Management Symposium (NOMS'98)*, vol. 3, pp. 733–742, March 1998.

APPENDIX A

Correctness of safe filter submission algorithm

In this appendix, we provide a correctness proof for the algorithm described in Figure 2.

Lemma 1: If f_1 and f_2 are active filters submitted by this algorithm, then f_1 and f_2 must have different filter names.

Proof: Notice that the maximum msg delivery is always less than the wait timer period. Without loss of generality, let us assume that f_1 was multicasted (line 3) at time t1 and f_2 was multicasted at time t2 where t1 $\leq$ t2. $WaitTime_1$ and $WaitTime_2$ are timers in line 4 and 9 respectively. It is one of these three cases:

(1) If $t2 - t1 > WaitTime_1$, then $Manager_2$ will abandon f_2 submission process because the name of the filter will be found when checking the MKB in line 2 of the algorithm.

(2) If $t2 - t1 \leq WaitTime_1$, then $Manager_2$ must receive the multicast notification of f_1 before the f_2 wait timer expires because the max arrival time of f_1 is less than f_2 waiting time ($t1 + WaitTime_1 \leq t_2 + WaitTime_2$. This implies that $Manager_1$ submits f_1 successfully (line 6) but $Manager_2$ finds f_1 name in its local MKB (line 5 in the algorithm). As a result of f_1 submission, f_1 is multicasted which causes $Manager_2$ to abandon the f_2 submission process (line 11) after $WaitTimer_2$ expires.

(3) If $t2 - t1 = 0$, then $Manager_1$ and $Manager_2$ must simultaneously receive f_1 and f_2 respectively before $WaitTimer_1$ expires. This causes both managers to find the filter name in the local MKB and the manager of the maximum IP address is

selected to submit this filter since no submission was performed before this point.

In all three possible cases, either f_1 or f_2 is allowed to be submitted exclusively because they have the same filter name. This means the assumption that f_1 and f_2 are successfully submitted into the system contradicts that f_1 and f_2 having the same name. The same proof is applied if $t1 > t2$.

Chapter 7

The ASIMUT simulation workshop

Monique Becker, Riadh Dhaou and Michel Marot
Institut National des Télécommunications, Evry, France

Andre-Luc Beylot
ENSEEIHT, Toulouse, France

Olivier Dalle and Philippe Mussi
INRIA, Sophia Antipolis, France

Vincent Sutter
CNES, Toulouse, France

Christian Rigal
Alcatel Space Industries, Toulouse, France

1. Objectives of the project

Performance evaluation is a crucial step in the process of designing and validating new satellite telecommunications systems. But performance evaluation of such systems is also a technical challenge. Out of the three classical methods for evaluating performances (analytical study, simulation study and experimental study) [JAI 91], only one may be applied in this context: the simulation one. Indeed, analytical studies can only cope with systems of a reasonable complexity. This is definitely not the case for the new generations of satellite telecommunications systems. And, as long as the whole system is not operational (this is obviously the case at the early stage of the design), experimental studies are either impossible, or restricted to small parts of the system.

Studies based on simulation techniques require both a computational model of the system and a simulation environment. Computational models are particularly difficult to elaborate because the system is large, complex and highly dynamic. At the physical level, for example, propagation effects are hard to model and often need to be accurately simulated in order to get realistic results. Furthermore, radio interfaces are now offering sophisticated mechanisms like power control or adaptive coding. Those particularities are added to the traditional complexity of multimedia networks based on ATM or IP networks and protocols. To face those new issues, a partnership between experts in the fields of multimedia network performance evaluation, propagation modeling, computer architecture and simulation, and satellite networks has been settled.

The team identified the following goals within the scope of the ASIMUT project [CON 00]:

– to identify the needs from the satellite manufacturing community in terms of simulation tools for performance evaluation;

– to design and develop a simulation environment introducing innovative features to ease model reuse and hierarchical modeling activities, as well as implementing powerful computing techniques;

– to define a generic satellite multimedia network model, and implement its simulation within the framework; all mechanisms shall be addressed in order to provide a global model of the system;

– to identify and evaluate appropriate simulation techniques in the satellite multimedia network context.

The generic network model, together with model reuse facilities offered by the tool, shall promote closer share and support amongst satellite telecommunication systems R&D community.

This paper presents the first results of this project. Section 2 emphasises on modeling problems that are specific to satellite multimedia networks simulation. Section 3 presents the original approach that will be implemented within ASIMUT simulation environment.

2. Satcom systems modeling

2.1. Modeling goals

There is a need for a dynamic and global simulator. It has to be dynamic because satellites are moving fast. It has to be global because a user moves from one satellite to another and it is necessary to take into account every satellite. This leads to the need for an extension of classical simulators, which were not designed for dynamic and global complex systems.

The goal is to evaluate the capacity, the availability, the quality of service, and the spectral effectiveness of satellite systems. The metrics used to estimate these parameters are the following:

Capacity is estimated by the maximum instantaneous traffic, the traffic mean, the total user count, as well as the average number of connections established during a given period.

Availability is evaluated from a probability of communication establishment, or a probability of communication maintenance.

Quality of service is evaluated at several levels, through the criteria of delay, jitter, BER, FER, ATM QoS parameters (CTD, CDV, CMR, CER, CLR, SECBR...), communication establishment overhead, connection outage time etc...

Evaluating the *spectral effectiveness* derived from an estimation of the supported traffic as a function of the number of subscribers, the geographical zone and service classes.

These studies concern the availability, effectiveness, the routing, the handoff, and the quality of service. We want to be able to simulate and compare several of these mechanisms and to integrate them in other studies. For example, a study of end-to-end QoS will integrate a handoff model but also a resource allocation model. We detail these levels later.

2.2. Main modeling problems

Models have to be designed on several levels and all the levels cannot be studied simultaneously. So an aggregation technique is necessary [MOU 98]. The complex system is shared into subsystems, which are studied independently. Then the global system is studied taking into account the subsystem dependencies. For example, to study a handoff mechanism, the failure problems are neglected and to study propagation problems, we do not have to simulate the whole constellation. Nevertheless, most of the mechanisms and phenomena are correlated and in several ways.

2.2.1. Cyclic dependence

A cyclic dependence occurs when several parts of the model, detailed in Figure 1, are correlated. For example, the performance of the radio channel depends on the traffic

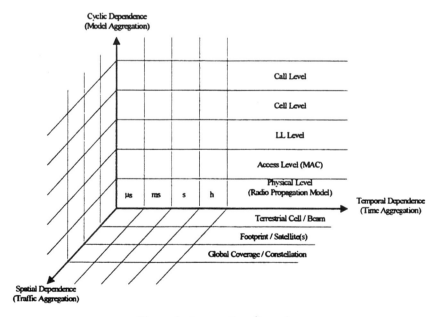

Figure 1. Aggregation dimensions

while the traffic itself depends on the performance of the radio channel, as shown in Figure 2. On the other side, the interference calculation utilizes the concurrent traffic. That leads to problems, for the representation of this concurrent traffic, for the relevance of its representation and for the importance to take into account.

In a satellite cell, many connections are handled simultaneously. Each of these connections generates a large amount of data, which results in a huge number of packets and a huge quantity of ATM cells. But for some studies, the communication needs to be simulated both at the network and radio levels, because they are closely dependent. However, it is nearly impossible to carry out simulations, simultaneously, on two different levels.

In this case, the problem we have to solve is the choice of the model smoothness. The performance of a simulation has a significant impact on its feasibility, and its complexity has a significant impact on the confidence intervals. Studies focusing on a given level in general require a relatively coarse description of the other levels. For example, for a study on the MAC layer level, an evaluation of the BER is required from the radio layer. But it is not necessarily useful to include detailed models of orbitography in these BER computations: approximating the movement of the satellites in a circle makes it possible not to have to integrate Kepler equations and to replace bulky files handling by simple recourse to mathematical functions.

2.2.2. Temporal dependence

The performance of the radio link at time t+1 depends on its performances at time t. This type of problem appears particularly when running step by step simulations with a very fine step and when running discrete event simulation with significantly correlated processes. That occurs when using detailed models of orbitography requiring, for their integration, a small step or when using long range dependence

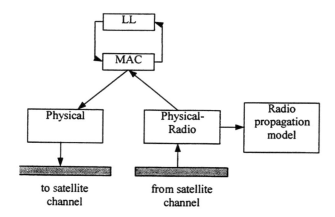

Figure 2. Cyclic dependence

traffic models. In most cases, a tradeoff has to be found between the smoothness of the model determining the step of the simulation and the desired result precision. Simulating process with long-term correlation is dangerous, because of the various time scales correlation. It is as dangerous as simulating non-stationary processes.

2.2.3. Space dependence

The performance on site S at time t is related to the performance on a close site S' at t or near time t. The co-localised space zones are dependent, for instance, because they share a handoff mechanism or another mechanism or phenomenon. The complexity of global studies taking into account all the system can be reduced by undertaking the study at an elementary space granularity and by using the results of this study to estimate the parameters on higher levels: iterative steps of traffic aggregation may be applied [MOU 98]. (See Figure 3.)

2.2.4. Choice of the granularity

The choice depends on the study topic. According to the type of study, various modeling levels of a same function or a same set of functions will be necessary. The modeling levels are represented in Figure 1. The multiple correlation is located; it is necessary here to identify these dependencies and treat them individually. In addition, some system variables (movements of the satellites) obey the regular laws and continuous variations. "Step-by-step" simulation is the more efficient one, in these cases. It is then necessary to combine this technique with "discrete events" simulations, more appropriate for simulating the behavior of telecommunication protocols. We have to find and validate some efficient mechanisms in order to combine different time, space or model scales.

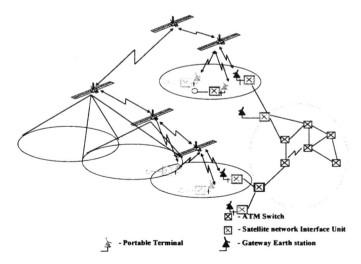

Figure 3. Space dependence

2.2.5. The problem of source models

The existing studies [MAR 00] on the terrestrial networks are difficult to project on the satellite networks, which do not exist yet. Moreover, the traffic models can evolve very quickly. It would be optimistic to consider only the traffic models generated by the users and applications of the current terrestrial networks. In addition, the user behavior varies according to the state of the underlying network, and, simultaneously, the applications adapt themselves to the characteristics of the networks.

Other factors specific to the satellite systems will modify the traffic nature. Contrary to current large scale networks, the considered architectures of satellite constellations are more adapted to some control mechanisms and present some geographical distribution characteristics and alternative use. (Obviously, it is false in the case of access constellations.)

However, taking into account the fact that the bandwidth of the terrestrial networks is continuously increasing and that these networks offer better throughput and delays than those offered by the satellite networks at their beginning, the user models on terrestrial networks can be taken as worst-cases.

Moreover, the traffic models depend on the user models. However, for services like WEB, except when the performance of the network is very bad, the user behavior is more dependent on the structure of the WEB site than on the performance of the network [MAR 00]. Indeed, the response time of the user is larger than the one of the network (the reading time of a WEB document is in general larger than its downloading), and thus the user model does not much depend on the network. In this context, it can be assumed that some traffic models, valid for the terrestrial networks, will also be valid for the LEO satellite networks (MEO and GEO are longer delays).

Most of the problems we have mentioned are not treated in traditional simulators [cf. section 3.1.1], which leaves the user to use some elementary block modules and to pile them up. But, the piled-up models cannot be simulated in a reasonable time to obtain realistic performance criteria. These problems are perhaps less crucial when local area networks or networks with a small number of nodes are simulated. Here the problem comes from the size (in terms of number of nodes or users) and from the diversity of the problems. It is thus advisable to consider them a priori to avoid disappointments.

Our view is that the only way to solve these problems is to provide for each object and modulus several visions that will depend on the level of necessary detail (time scale and space scale).

3. ASIMUT: a new environment for SATCOM systems modeling

The motivation for designing a completely novel simulation environment results from the observation that the usual approaches for studying multimedia satellite telecommunications systems are laborious and inefficient. In paragraph 3.1, we summarise and discuss these usual approaches. Then, in paragraph 3.2, we describe the solution we propose with the ASIMUT simulation workshop.

3.1. Usual approaches

When the need for simulating a new satellite telecommunications system arises, the two usual approaches are either to build a new model of the targeted system on top of an existing simulation environment, or to build a new and therefore proprietary simulation software, specifically designed for a given system. In the following, we give a short analysis of the advantages and drawbacks of these two approaches.

3.1.1. Reusing existing simulation environment

Reusing an existing simulation environment, such as OPNET [OPN 00], STK [STK 00], NS [NS 00], Visualyze, COSSAP or SPW [SPW 92], has several advantages. First, existing environments usually include a set of common models or patterns out of which several may be reused. Second, the larger the user community of an environment, the easier it is to find support and contributed models for this environment. Third, existing environments often come with a set of integrated or contributed tools, such as model animators and debuggers, plotters, data analyzers, and so on, which improves their overall ergonomics and efficiency.

Unfortunately, satellite telecommunications systems are combining several aspects that used to be studied separately. Therefore, these separate kinds of studies lead to specific environments. Some are more specifically designed for network and protocol modeling, others for propagation and radio interference modeling, and others for space mechanics modeling. Of course, out of these specific environments, some have the ability to be extended to new areas. But since a specific environment usually means a specific and optimized design, such an extension is seldom easy: integrating new kinds of models often conflicts with the initial design and philosophy of the selected environment. The result is an added modeling complexity, as well as an added computing complexity, which are both critical points given the high complexity and large scale of the systems being studied.

3.1.2. Building specific simulation software

Compared to the previous approach, building a specific simulation software has the opposite advantages and drawbacks: the modeling complexity may be lowered to a minimal level and the computing complexity may be sharply optimized. But since the developments are specific and often proprietary, there is nothing or little to share, no community support, and a lot of additional effort is required to develop specific complementary tools or integrate existing ones.

3.2. The ASIMUT environment

ASIMUT[1] is a workshop that provides, through an integrated user interface, all the functions required to achieve an experimental study based on simulation: model programming and assembly, experiment planning, simulation runtime support, and data analysis.

3.2.1. General architecture and concepts

The general architecture and main concepts of the ASIMUT workshop are depicted in Figure 4. The workshop front-end is a user interface that addresses three categories of users: developers, model architects and experimenters. Developers are in charge of implementing the code and algorithms of the basic elements of the models, such as queues, protocols, radio links, propagation models or traffic generators. Model architects use these basic elements to build more complex components, such as protocol stacks, satellites, terminal, gateways or simulation dynamics. The architects then use all these elements to build complete system models and make them available in the reference space. Experimenters

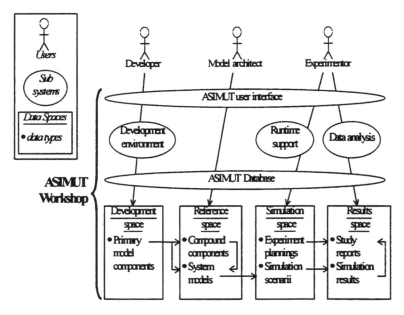

Figure 4. General architecture and main concepts of the ASIMUT workshop

[1] Atelier de SImulation de systèmes MUltimedia par satelliTes: Workshop for the simulation of multimedia systems over satellites.

build their simulation scenarios by first selecting the most appropriate system model in the reference space. This model is then used to derive a set of simulation scenarios. Each scenario is an instance of the system model, in which all the parameters of the model are given a specific value. These specific values are chosen according to a global policy, in order to optimise the number of simulation runs required for the study. The experimenter establishes this policy by defining an experiment plan. Once all the simulation runs are complete, the experimenter enters the final stage of analysing the data collected. This analysis leads to study reports, which exhibit the relations that exist between the model parameters chosen through the experiment plan and the data collected during the runs.

The introduction of a representative physical layer model under the network components is a key issue in the project. Very heterogeneous radio links will be simulated. There will be inter-satellite links, mobile environment or high frequency gateway links. All those links must rely on models that allow the experimenter to assess their impact on cell delay variations, interference level variation due to traffic load and meteorological events. A classical approach rapidly falls into granularity problems because of the transmission rates diversity and the number of links that must be simulated concurrently. To overcome this problem, a specific packetization approach has been introduced. It is based on a simple segmentation and aggregation process that allows all links to work with similar time frames, not directly related to the transported stream bit rate. This flexibility opens a wide area of modeling possibilities as the radio layer behavior will only be sampled at a rate corresponding to its physical characteristics, independently from the traffic stream.

3.2.2. Innovative elements of design

SATCOM systems exhibit several levels of complexity: a complexity at a physical level with the radio transmissions, a complexity at an architectural level, with several kinds of nodes (terminal, satellites, gateways, control centers) and links, a complexity at a scale level, with networks operating simultaneously up to thousands of terminals, and a complexity at a functional level, with several kinds of services (e.g.: ATM-like services), several kinds of protocols (e.g.: IP and ATM stacks) and several kinds of procedures (e.g.: logon, power-control, CAC, hand-off). In order to cope with all these levels of complexity, the ASIMUT workshop implements and provides several innovative techniques.

3.2.2.1. Component-based modeling approach

The component-based modeling approach is a key point in the ASIMUT design. This common pattern of object oriented programming [GAM 94], applied to the modeling area, allows for many interesting features, such as model genericity, variable granularity, and hierarchical modeling.

Genericity is a powerful property of ASIMUT models that allows the transparent reuse or exchange of any part of a given model in order to build new models. Variable granularity is the ability to switch the detail level of a given part of a model, from the most detailed level to the most approximate level. Hierarchical modeling is the ability to decompose complex parts of a model in simpler parts, until a reasonable complexity level is reached. An aggregation technique may then be used.

3.2.2.2. Open and versatile simulation kernel

ASIMUT simulator architecture is based on a framework approach. This will allow for maximal flexibility and adaptability at all levels.

Typical complex simulators have shown to spend most of the computing time in event handling. Thus, this part of the simulator has to be carefully designed and optimized. But in the case of SATCOM systems, the kind of designs and optimizations required depends on the part of the model being simulated. In order to allow several kinds of optimization to be combined, ASIMUT kernel will allow for transparent mixed-mode multi-level kernels:

– continuous and clock-based parts of the models, such as propagation and radio link models, will benefit from a time-step kernel mode;
– high level or irregular parts, such as protocols, will run on an event-driven kernel mode;
– multiple kernels (possibly of different types) will be able to run concurrently for the same simulation, under the control of a higher level kernel.

This multi-level execution architecture will also facilitate the introduction of distribution, and the efficient use of multi-processor or clustered platforms. Let us note, that transparent (at user-level) and efficient use of distributed simulation has not been really achieved for such complex systems. However, one can reasonably hope that high performance distributed subsystems may be implemented by means of special-purpose libraries.

3.2.2.3. Dynamicity

Dynamicity is the ability to create and destroy model components during a simulation run. For example, this property is required in order to create and destroy radio links components. But it also appears to be useful in order to cope with the model complexity. For example, it may be useful to dynamically aggregate and split clusters of terminals, depending for example on their relative positions to a given low earth orbital satellite.

4. Status of the project

The ASIMUT project is a three-year effort that started by the end of 1999. This effort was divided into two main tasks: (i) design and develop a new simulation environment and (ii) build advanced models of satellite telecommunications systems for this environment.

The first task is divided into three stages: study simulation techniques and select the most relevant ones for the implementation in the ASIMUT environment [BEC 99]; produce technical software specifications of the simulation environment [DAL 00, DAL 01]; and develop the simulation environment according to previous specifications. The first task has just entered the third stage and is expected to complete soon with the delivery of ASIMUT simulation environment (all the software developments are conducted by a third-party company). The ASIMUT simulation environment will provide a GUI based on the Java technology. It will embed Prosit C++ framework, a high performance discrete-event simulation tool developed at INRIA [SIE 97].

The second task is also devided into three stages: design satellite networks models based on the new simulation environment capabilities; write the corresponding technical specifications; and develop the corresponding models. This task, which requires input from the first task, has just started; it is expected to complete by the end of 2002, with the delivery of a full set of generic model components. This set of components will allow a designer to model a typical satellite telecommunications system: a Ka band cellular system with adaptive coding and on-board (cell) switching, MF-TDMA or CDMA access, dynamic resource allocation (CF-DAMA) and QoS support for typical multimedia applications.

Besides these two main tasks, additional studies on simulation and modeling techniques [BEC 99] are still being conducted by the project members during the development of the ASIMUT simulation workshop. Still, a major issue will be to determine more precisely the accurate granularity of the models depending on the kind of analysis to be performed. Important experimentation and validation phases will be expected for that purpose.

REFERENCES

[BEC 99] BECKER M., BEYLOT A.L., MAROT M., "Etat de l'Art de la Simulation et des Outils de Simulation (Survey of Simulation and Simulation Tools)". Delivered at *Multimedia Satellite Constellation Project, Convention*, 5 June 1999.

[CER 99] CERCAS F., "Simulation Tools", *COST253*, May 1999.

[CON 00] The Multimedia Satellite Constellation Grant. http://constellation.prism.uvsq.fr/

[DAL 00] DALLE O., SUTTER V., RIGAL C., "Spécification Technique de Besoin Logiciel de l'Environnement ASIMUT (Technical Software Specification of the ASIMUT Environment)", *Specification ATF-SB-1-00–29-CNES*, CNES, March 2000.

[DAL 01] DALLE O., RIGAL C., "Exemple d'Utilisation de l'Environnement ASIMUT (An Example of the ASIMUT Environment Usage)", *Technical report ATF-SB-1-00–43-CNES*, CNES, March 2000.

[GAM 94] GAMMA E., HELM R., JOHNSON R., VLISSIDES J., *Design Patterns: Elements of Reusable Object-Oriented Software*, Addison Wesley, Massachusetts, 1994.

[JAI 91] JAIN R., *The Art of Computer Performance Analysis: Techniques for Experimental Design, Measurement, Simulation and Modelling*, Wiley, 1991.

[MAR 00] MAROT M., BECKER M., VINCENT P., "Modeling a WEB User, Application to Performance Comparison of UDP Versus TCP for WEB Traffic Transport", *ICTS 2000*, 2000.

[MOU 98] MOULKI M., BEYLOT A.L., TRUFFET L., BECKER M., "An Aggregation Technique to Evaluate the Performance of a Two-stage Buffered ATM Switch", *Annals of Operations Research*, n° 79, 1998, p. 373–392.

[NS 00] NS, Network Simulator. http://www-mash.cs.berkeley.edu/ns/

[OPN 00] OPNET™, OPNET Simulator. MIL3 product, http://www.mil3.com

[SIE 97] SIEGEL G., "Prosit: un Environnement pour la Programmation de Simulations à Événements Discrets (Prosit: an Environment for Discrete Event Simulations Programming)", PhD thesis, INRIA/University of Nice, Sophia Antipolis, September 1997.

[SPW 92] SPW™, Signal Processing Work-System, "The DSP Framework Users Guide and Tutorial", COMDISCO Systems, Inc. September 1992.

[STK 00] STK™, Satellite ToolKit. http://www.stk.com

Chapter 8

Closed set-based discovery of small covers for association rules

Nicolas Pasquier
I3S – CNRS UPRESA 6070 – UNSA, Les Algorithmes-Euclide B, Sophia Antipolis, France

Yves Bastide and Rafik Taouil
LIMOS – CNRS FRE 2239, Université Blaise Pascal, Aubière, France

Lotfi Lakhal
LIM – CNRS FRE 2246, Université de la Méditérranée, Marseille, France

1. Introduction and motivation

Data mining has been extensively addressed over recent years, especially the problem of discovering association rules. The aim when developing association rules is to exhibit relationships between data items (or attributes) and compute the precision of each relationship in the database. The customary precision measures used are support and confidence [AGR 93] that point to the proportion of database transactions (or objects) upholding each rule. When an association rule has support and confidence exceeding some user-defined minimum threshold, the rule is considered to be relevant and the extracted knowledge would likely be used for supporting decision making. A classical example of association rules fits in the context of market basket data analysis and highlights a particular feature in customer behavior: 80% of customers who buy cereals and sugar also buy milk and 20% of customers buy all three items.

Since the problem was first stated [AGR 93], various approaches have been proposed to increase efficiency of rule discovery [AGR 94, BAY 98, BRI 97b, LIN 98, PAS 98, PAS 99b, PAS 99a, SAV 95, TOI 96, ZAK 97]. However, taking full advantage of exhibited knowledge means capabilities to handle such knowledge. In fact, by using a synthetic dataset containing 100,000 objects, each of which encompassing around 10 items, our experiments yield more than 16,000 rules with a confidence outcome of 90%. The problem is much more critical when collected data is highly correlated or dense, like in statistical or medical databases. For instance, when applied to a census dataset of 10,000 objects, each of which is characterized by values of 73 attributes, experiments result in more than 2,000,000 rules with a support and confidence outcome of 90%.

So the issue under discussion follows: which relevant knowledge can be learned from several thousands of highly redundant rules? What aid could be offered to users for handling countless rules and focusing on useful ones? Before explaining how our approach answers the previous questions, let us examine the proposed solutions for meeting such needs.

1.1. Related work: an outline

Among approaches addressing this issue, two main trends can be distinguished. The former provides users with mechanisms for filtering rules. In [BAR 97, KLE 94], the user defines templates, and rules not matching them are discarded. In [NG 98, SRI 97], Boolean operators are introduced for selecting rules including (or not) given items. In [SRI 96, TOI 95], methods for pruning rules with weak measures of improvement that characterize the difference between supports and confidences of a rule and its sub-rules, i.e. with smaller antecedent and same consequent, are proposed. A similar approach expanded by Boolean operators for selecting rules is proposed in [BAY 99b]. In [MEO 96], an SQL-like operator called MINE RULE, allowing the specification of general extraction criteria, is defined. The use of the user's domain knowledge for selecting unexpected rules, using measures of distance between rules called deviation measures, is proposed in [HEC 96, PIA 91, SIL 96]. In [BAY 99a], the proposed approach consists in selecting rules with maximal antecedent, called A-maximal rules, which are rules for which the addition of an item to the antecedent reduces the population governed by the rule. The quoted approaches operate "a posteriori", i.e. once huge numbers of rules are extracted, querying facilities make it possible to handle rule subsets selected according to the user preferences.

In contrast, the second trend addresses the problem with an "a priori" vision, by attempting to minimize the number of exhibited rules. In [HAN 95, SRI 95], information about taxonomies is used to define criteria of interest which apply for pruning redundant rules. The use of statistical measures, such as Pearson's correlation, chi-squared test, conviction, interest, entropy gain, gini or lift, instead of the confidence measure is studied in [BRI 97a, MOR 98, SIL 98].

1.2. Contribution: an overview

The approach presented in this paper belongs to the second trend since it aims to extract not all possible rules but a subset called *small* cover or *basis* for association rules. When computing such a basis, redundant rules are discarded since they do not carry relevant knowledge. Such a pruning operation is a key step for rule extraction, and significantly reduces the resulting set. Moreover, since rules that are unexpected by the user are important [LIU 97, SIL 96], presentation of a list of rules covering all the frequent items in the dataset is also needed. The approach proposed in this paper meets this requirement.

First, using the closure operator of the Galois connection [BIR 67], we characterize frequent closed itemsets introduced in [PAS 98]. Then, we show that frequent closed itemsets represent a generating set for both frequent itemsets and association rules. The underlying theorem provides the foundations of our approach since it makes it possible to generate the bases from frequent closed itemsets by avoiding handling of large sets of rules. We propose two new algorithms: the former achieves frequent closed itemsets from frequent itemsets without accessing the dataset, and the latter, called Apriori-Close, extends the Apriori algorithm [AGR 94] by discovering simultaneously frequent itemsets and frequent closed itemsets without additional execution time.

Then, using the frequent closed itemsets and the pseudo-closed itemsets defined by Duquenne and Guigues in lattice theory [BUR 98, DUQ 86], we define the *Duquenne-Guigues basis for exact association rules* (rules with a 100% confidence). Rules in this basis are non-redundant exact rules. Besides, using the frequent closed itemsets and results proposed by Luxenburger in lattice theory [LUX 91], we define the *proper basis* and the *structural basis for approximate association rules*. The proper basis is a small set containing non-redundant approximate association rules. The structural basis can be viewed as an abstract of all approximate rules that hold and can be useful when the proper basis is large. We propose three algorithms intended for yielding these three bases. Using the set of frequent closed itemsets, generating the evoked bases is performed without any access to the dataset.

An algorithm discovering closed and pseudo-closed itemsets has been proposed in [GAN 91] and implemented in CONIMP [BUR 98]. However, this algorithm does not consider the support of itemsets and, since it works only in main memory, it cannot be applied when the number of objects exceeds some hundreds and the number of items some tens. From the results presented in [LUX 91], no algorithm was proposed. In [PAS 98, PAS 99a], the association rule framework based on the Galois connection is defined. Fitting in this groundwork, two efficient algorithms that discover frequent closed itemsets for association rules are defined: the Close algorithm [PAS 98, PAS 99a] for correlated data and the A-Close algorithm [PAS 99b] for weakly correlated data. The work presented in this paper differs from [PAS 98, PAS 99a, PAS 99b] in the following ways:

1. It shows that frequent closed itemsets constitute a generating set for frequent itemsets and association rules.

2. It extends the Apriori algorithm and algorithms for discovering maximal frequent itemsets to generate frequent closed itemsets.

3. It adapts the Duquenne-Guigues basis and Luxenburger results for exact and partial implications to the context of association rules. This adaptation is based on 1 (generating set).

4. It presents new algorithms for generating bases for exact and approximate association rules using frequent closed itemsets.

5. It shows that the algorithms proposed are efficient for both improving the usefulness of extracted association rules and decreasing the execution time of the association rule extraction.

1.3. Paper organization

In Section 2, we present the association rule framework based on the Galois connection. Section 3 addresses the concept of basis for both exact and approximate association rules. New algorithms for discovering frequent and frequent closed itemsets are described in Section 4 and the following section presents algorithms computing the bases for association rules from the frequent closed itemsets. Experimental results achieved from various datasets are given in Section 6. Finally, as a conclusion, we evoke further work in Section 7.

2. Association rule framework

In this section, we present the association rule framework based on the Galois connection, primarily introduced in [PAS 98].

Definition 1 (Data mining context). *A data mining context[1] is defined as* $\mathcal{D} = (O, \mathcal{1}, \mathcal{R})$, *where* O *and* $\mathcal{1}$ *are finite sets of objects and items respectively.* $\mathcal{R} \subseteq O \times \mathcal{1}$ *is a binary relation between objects and items. Each couple* $(o, i) \in \mathcal{R}$ *denotes the fact that the object* $o \in O$ *is related to the item* $i \in \mathcal{1}$.

Depending on the target system, a data mining context can be a relation, a class, or the result of an SQL/OQL query.

Example 1. An example data mining context $\mathcal{D}$ consisting of 5 objects (identified by their OID) and 5 items is illustrated in Table 1.

Table 1. The example data mining context $\mathcal{D}$

OID	Items			
1	A	C	D	
2	B	C	E	
3	A	B	C	E
4	B	E		
5	A	B	C	E

Definition 2 (Galois connection). *Let* $\mathcal{D} = (O, \mathcal{1}, \mathcal{R})$ *be a data mining context. For* $O \subseteq O$ *and* $I \subseteq \mathcal{1}$, *we define:*

$$f : 2^O \to 2^{\mathcal{1}} \qquad\qquad g : 2^{\mathcal{1}} \to 2^O$$
$$f(O) = \{i \in \mathcal{1} \mid \forall o \in O, (o, i) \in \mathcal{R}\} \qquad g(I) = \{o \in O \mid \forall i \in I, (o, i) \in \mathcal{R}\}$$

---

[1] By extension, we will call dataset a data mining context.

f (O) associates with O the items common to all objects $o \in O$ *and g (I) associates with I the objects related to all items* $i \in I$. *The couple of applications (f, g) is a Galois connection between the power set of* O (2^O) *and the power set of* $\mathcal{I}$ (2^I). *The following properties hold for all I,* I_1, $I_2 \subseteq \mathcal{I}$ *and O,* O_1, $O_2 \subseteq O$:

$$(1)\ I_1 \subseteq I_2 \Rightarrow g(I_1) \supseteq g(I_2) \qquad\qquad (1')\ O_1 \subseteq O_2 \Rightarrow f(O_1) \supseteq f(O_2)$$

$$(2)\ O \subseteq g(I) \Leftrightarrow I \subseteq f(O)$$

Definition 3 (Frequent itemsets). *Let* $I \subseteq \mathcal{I}$ *be a set of items from* $\mathcal{D}$. *The support count of the itemset I in* $\mathcal{D}$ *is:*

$$supp(I) = \frac{|g(I)|}{|O|}$$

I is said to be frequent if the support of I in $\mathcal{D}$ *is at least minsupp. The set L of frequent itemsets in* $\mathcal{D}$ *is:*

$$L = \{I \subseteq \mathcal{I} \mid supp(I) \geq minsupp\}$$

Definition 4 (Association rules). *An association rule is an implication between two itemsets, with the form* $I_1 \rightarrow I_2$ *where* I_1, $I_2 \subseteq \mathcal{I}$, I_1, $I_2 \neq \emptyset$ *and* $I_1 \cap I_2 = \emptyset$. I_1 *and* I_2 *are called respectively the antecedent and the consequent of the rule. The support supp(r) and confidence conf(r) of an association rule r:* $I_1 \rightarrow I_2$ *are defined using the Galois connection as follows:*

$$supp(r) = \frac{|g(I_1 \cup I_2)|}{|O|}, \quad conf(r) = \frac{supp(I_1 \cup I_2)}{supp(I_1)}$$

Association rules holding in this context are those that have support and confidence greater than or equal to the minsupp and minconf thresholds respectively. We define the set AR of association rules holding in $\mathcal{D}$ *given minsupp and minconf thresholds as follows:*

$$AR = \{r : I_1 \rightarrow I_2 \setminus I_1 \mid I_1 \subset I_2 \subseteq \mathcal{I} \wedge supp(I_2) \geq minsupp$$
$$\wedge\ conf(r) \geq minconf\}$$

If conf(r)=1 then r is called an exact association rule or implication rule, otherwise r is called approximate association rule.

Example 2. Exact and approximate association rules extracted from $\mathcal{D}$ for *minsupp* = 2/5 and *minconf* = 1/2 are given in Table 2.

Table 2. *Association rules extracted from $\mathcal{D}$ for minsup = 2/5 and minconf = 1/2*

Exact rule	Supp	Approximate rule	Supp	Conf	Approximate rule	Supp	Conf
ABC ⇒ E	2/5	BCE → A	2/5	2/3	B → AE	2/5	2/4
ABE ⇒ C	2/5	AC → BE	2/5	2/3	E → AB	2/5	2/4
ACE ⇒ B	2/5	BE → AC	2/5	2/4	A → CE	2/5	2/3
AB ⇒ CE	2/5	CE → AB	2/5	2/3	C → AE	2/5	2/4
AE ⇒ BC	2/5	AC → B	2/5	2/3	E → AC	2/5	2/4
AB ⇒ C	2/5	BC → A	2/5	2/3	B → CE	3/5	3/4
AB ⇒ E	2/5	BE → A	2/5	2/4	C → BE	3/5	3/4
AE ⇒ B	2/5	AC → E	2/5	2/3	E → BC	3/5	3/4
AE ⇒ C	2/5	CE → A	2/5	2/3	A → B	2/5	2/3
BC ⇒ E	3/5	BE → C	3/5	3/4	B → A	2/5	2/4
CE ⇒ B	3/5	A → BCE	2/5	2/3	C → A	3/5	3/4
A ⇒ C	3/5	B → ACE	2/5	2/4	A → E	2/5	2/3
B ⇒ E	4/5	C → ABE	2/5	2/4	E → A	2/5	2/4
E ⇒ B	4/5	E → ABC	2/5	2/4	B → C	3/5	3/4
		A → BC	2/5	2/3	C → B	3/5	3/4
		B → AC	2/5	2/4	C → E	3/5	3/4
		C → AB	2/5	2/4	E → C	3/5	3/4
		A → BE	2/5	2/3			

3. Bases for association rules

In this section, we first demonstrate that the frequent closed itemsets constitute a generating set for frequent itemsets and association rules. Then, we characterize the *Duquenne-Guigues basis for exact association rules* and the *proper* and *structural bases for approximate association rules*. These bases are adaptions of the bases defined by Duquenne and Guigues [DUQ 86] and Luxenburger [LUX 91] in Lattice Theory and Data Analysis to the context of association rules. This adaptation is not trivial since additional constraints related to the specificity of association rules have to be considered. Theorem 2 states that the union of the Duquenne-Guigues basis for exact association rules and the proper basis or the structural basis for approximate association rules constitutes a basis for all valid association rules. The proof of this theorem is straightforward from Theorem 1 and [DUQ 86, LUX 91]. Interested readers could refer to [BIR 67, GAN 99, WIL 92] for further details on closed sets.

3.1. Generating set

Definition 5 (Galois closure operators). *The operators $h = fog$ in $2^{\mathcal{I}}$ and $h' = gof$ in 2^O are Galois closure operators[2]. Given the Galois connection (f, g), the following properties hold for all I, I_1, $I_2 \subseteq \mathcal{I}$ and O, O_1, $O_2 \subseteq O$ [BIR 67]:*

---

[2] Here, we use the following notation: $fog(I) = f(g(I))$ and $gof(O) = g(f(O))$.

Extension: $(3)\ I \subseteq h(I)$ $(3')\ O \subseteq h'(O)$
Idempotency: $(4)\ h(h(I) = h(I)$ $(4')\ h'(h'(O)) = h'(O)$
Monotonicity: $(5)\ I_1 \subseteq I_2 \Rightarrow h(I_1) \subseteq h(I_2)$ $(5')\ O_1 \subseteq O_2 \Rightarrow h'(O_1) \subseteq h'(O_2)$

Definition 6 (Frequent closed itemsets). *An itemset $I \subseteq 1$ in $\mathcal{D}$ is a closed itemset iff $h(I) = I$. A closed itemset I is said to be frequent if the support of I in $\mathcal{D}$ is at least minsupp. The smallest (minimal) closed itemset containing an itemset I is h(I), the closure of I. The set FC of frequent closed itemsets in $\mathcal{D}$ is defined as follows:*

$$FC = \{I \subseteq 1 \mid I = h(I) \land supp(I) \geq minsupp\}$$

Example 3. A frequent closed itemset is a maximal set of items common to a set of objects, for which support is at least *minsupp*. The frequent closed itemsets in the context $\mathcal{D}$ for *minsupp* = 2/5 are presented in Table 3. The itemset *BCE* is a frequent closed itemset since it is the maximal set of items common to the objects $\{2, 3, 5\}$. The itemset *BC* is not a frequent closed itemset since it is not a maximal set of items common to some objects: all objects in relation with the items B and C (objects 2, 3 and 5) are also in relation with the item E.

Hereafter, we demonstrate that the set of frequent closed itemsets with their support is the smallest collection from which frequent itemsets with their support and association rules can be generated (it is a generating set).

Table 3. *Frequent closed itemsets extracted from $\mathcal{D}$ for minsupp = 2/5*

Frequent closed itemset	Support
$\{\varnothing\}$	5/5
$\{C\}$	4/5
$\{AC\}$	3/5
$\{BE\}$	4/5
$\{BCE\}$	3/5
$\{ABCE\}$	2/5

Lemma 1. *[PAS 98, PAS 99a] The support of an itemset I is equal to the support of the smallest closed itemset containing I: $supp(I) = supp(h(I))$.*

Lemma 2. *[PAS 98, PAS 99a] The set of maximal frequent itemsets $M = \{I \in L \mid \nexists\ I' \in L$ where $I \subset I'\}$ is identical to the set of maximal frequent closed itemsets $MC = \{I \in FC \mid \nexists\ I' \in FC$ where $I \subset I'\}$.*

Theorem 1 (Generating set). *The set FC of frequent closed itemsets with their support is a generating set for all frequent itemsets and their support, and for all association rules holding in the dataset, their support and their confidence.*

Proof. Based on Lemma 2, all frequent itemsets can be derived from the maximal frequent closed itemsets. Based on Lemma 1, the support of each frequent itemset can be derived from the support of frequent closed itemsets. Then, the set of frequent closed itemsets FC is a generating set for both the set of frequent itemsets L and the set of association rules AR[3].

3.2. Duquenne-Guigues basis for exact association rules

Definition 7 (Frequent pseudo-closed itemsets). *An itemset $I \subseteq 1$ in $\mathcal{D}$ is a pseudo-closed itemset iff $h(I) \neq I$ and $\forall I' \subset I$ such that I' is a pseudo-closed itemset, we have $h(I') \subset I$. The set FP of frequent pseudo-closed itemsets in $\mathcal{D}$ is defined as*

$$FP = \{I \subseteq 1 \mid supp(I) \geq minsupp \ \wedge \ I \neq h(I) \ \wedge \ \forall I' \in FP \text{ such that } I' \subset I \\ \text{we have } h(I') \subset I\}$$

Definition 8 (Duquenne-Guigues basis for exact association rules). *Let FP be the set of frequent pseudo-closed itemsets in $\mathcal{D}$. The Duquenne-Guigues basis for exact association rules is defined as:*

$$DG = \{r : I_1 \Rightarrow h(I_1) \setminus I_1 \mid I_1 \in FP \wedge I_1 \neq \varnothing\}$$

The Duquenne-Guigues basis is minimal with respect to the number of rules since there can be no complete set with fewer rules than there are frequent pseudo-closed itemsets [DEM 92, GAN 99].

Example 4. A frequent pseudo-closed itemset I is a frequent non-closed itemset that includes the closures of all frequent pseudo-closed itemsets included in I. The set FP of frequent pseudo-closed itemsets and the Duquenne-Guigues basis for exact association rules extracted from $\mathcal{D}$ for $minsupp = 2/5$ and $minconf = 1/2$ are presented in Table 4. The itemset AB is not a frequent pseudo-closed itemset since the closures of A and B (respectively AC and BE) are not included in AB. $ABCE$ is not a frequent pseudo-closed itemset since it is closed.

Table 4. *Frequent pseudo-closed itemsets and Duquenne-Guigues basis extracted from $\mathcal{D}$ for $minsupp = 2/5$*

Frequent pseudo-closed itemset	Support	Exact rule	Support
{A}	3/5	A ⇒ C	3/5
{B}	4/5	B ⇒ E	4/5
{E}	4/5	E ⇒ B	4/5

---

[3] Furthermore, FC is the smallest generating set for L and AR. Hence, even if frequent itemsets can be derived from the maximal frequent itemsets, passes over the dataset are still needed to compute the frequent itemset supports.

3.3. Proper basis for approximate association rules

Definition 9 (Proper basis for approximate association rules). *Let FC be the set of frequent closed itemsets in* $\mathcal{D}$. *The proper basis for approximate association rules is:*

$$PB = \{r : I_1 \rightarrow I_2 \setminus I_1 \mid I_1, I_2 \in FC \wedge I_1 \neq \emptyset \wedge I_1 \subset I_2 \wedge conf(r) \geq minconf\}$$

Association rules in PB are proper approximate association rules.

Example 5. The proper basis for approximate association rules extracted from $\mathcal{D}$ for *minsupp* = 2/5 and *minconf* = 1/2 is presented in Table 5.

Table 5. *Proper basis extracted from* $\mathcal{D}$ *for minsupp = 2/5 and minconf = 1/2*

Approximate rule	Support	Confidence
BCE → A	2/5	2/3
AC → BE	2/5	2/3
BE → AC	2/5	2/4
BE → C	3/5	3/4
C → ABE	2/5	2/4
C → BE	3/5	3/4
C → A	3/5	3/4

3.4. Structural basis for approximate association rules

Definition 10 (Undirected graph $\mathcal{G}_{FC}$**).** *Let FC be the set of frequent closed itemsets in* $\mathcal{D}$. *We define* $\mathcal{G}_{FC} = (V, E)$ *as the undirected graph associated with FC where the set of vertices V and the set of edges E are defined as follows:*

$$V = \{I \subseteq \mathcal{I} \mid I \in FC\}$$
$$E = \{(I_1, I_2) \in V \times V \mid I_1 \subset I_2 \wedge supp(I_2)/supp(I_1) \geq minconf\}$$

With each edge in $\mathcal{G}_{FC}$ *between two vertices* I_1 *and* I_2 *with* $I_1 \subset I_2$ *is associated the confidence = supp* (I_2) */ supp* (I_1) *of the proper approximate association rule* $I_1 \rightarrow I_2 \setminus I_1$ *represented by the edge.*

Definition 11 (Maximal confidence spanning forest $\mathcal{F}_{FC}$**).** *Let* $\mathcal{F}_{FC} = (V, E')$ *be the maximal confidence spanning forest associated with FC.* $\mathcal{F}_{FC}$ *is obtained from the undirected graph* $\mathcal{G}_{FC} = (V, E)$ *by suppressing transitive edges and cycles. Cycles are removed by deleting some edges that enter the last vertex I (maximal vertex with respect to the inclusion) of the cycle. Among all edges entering in I, those with confidence less than the maximal confidence value associated with an edge with the form* $(I', I) \in E$ *are deleted. If more than one edge has the maximal confidence value, the first one in lexicographic order is kept. (See Figure 1.)*

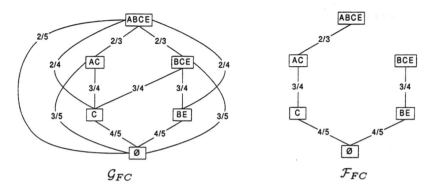

Figure 1. *Undirected graph $\mathcal{G}_{FC}$ and maximal confidence spanning forest $\mathcal{F}_{FC}$ (a tree in this example) derived from $\mathcal{D}$ for minsupp = 2/5 and minconf = 1/2*

Definition 12 (Structural basis for approximate association rules). *Let SB be the set of association rules represented by edges in $\mathcal{F}_{FC}$ except rules from the vertex $\{\varnothing\}$. The structural basis for approximate association rules is:*

$$SB = \{r : I_1 \to I_2 \setminus I_1 \mid I_1, I_2 \in V \wedge I_1 \subset I_2 \wedge I_1 \neq \varnothing \wedge (I_1, I_2) \in E'\}$$

In this basis, each frequent closed itemset is the consequent of at most one approximate association rule.

Example 6. The structural basis for approximate association rules extracted from $\mathcal{D}$ for *minsupp* = 2/5 and *minconf* = 1/2 is presented in Table 6.

Table 6. *Structural basis extracted from D for minsupp = 2/5 and minconf = 1/2*

Approximate rule	Support	Confidence
AC → BE	2/5	2/3
BE → C	3/5	3/4
C → A	3/5	3/4

3.5. Basis for all valid association rules

Theorem 2 (Basis for valid association rules). *The union of the Duquenne-Guigues basis for exact association rules and the proper basis or the structural basis for approximate association rules is a basis for all valid association rules, their support and their confidence.*

Proof. The proof of this theorem is straightforward from Theorem 1 and results presented in [DUQ 86, LUX 91]. All frequent closed itemsets and their support can be derived from the union of the Duquenne-Guigues basis and the proper or the structural basis since for each frequent closed itemset I_2 there exists at least one association rule of the form $r: I_1 \rightarrow I_2 \setminus I_1$ with $supp(I_2) = supp(r)$ and $supp(I_1) = supp(r)/conf(r)$. Moreover, all valid association rules can be derived with their support from the set FC of frequent closed itemsets (Theorem 1). Then, obviously, all valid association rules, their support and their confidence can be derived from this union.

4. Discovering frequent and frequent closed itemsets

In Section 4.1, we propose a new algorithm to achieve frequent closed itemsets from frequent itemsets without accessing the dataset. This algorithm discovers frequent closed itemsets while for instance an algorithm for discovering maximal frequent itemsets [BAY 98, LIN 98, ZAK 97] is used. In Section 4.2, we present an extension of the Apriori algorithm [AGR 94] called Apriori-Close for discovering frequent and frequent closed itemsets without additional computation time. Like in the Apriori algorithm, we assume in the following that items are sorted in lexicographic order and that k is the size of the largest frequent itemsets. Based on Lemma 2, k is also the size of the largest frequent closed itemsets.

4.1. Computing frequent closed itemsets from frequent itemsets

Many efficient algorithms for mining frequent itemsets and their support have been proposed. Well-known proposals are presented in [AGR 94, BRI 97b, SAV 95, TOI 96]. Efficient algorithms for discovering the maximal frequent itemsets and then achieving all frequent itemsets have also been proposed [BAY 98, LIN 98, ZAK 97]. All these algorithms give as a result the set $L = \cup_{i=1}^{i=k} L_i$ where L_i contains all frequent i-itemsets (itemsets of size i). Based on Proposition 1 and Lemma 2 (Section 3.1), the frequent closed itemsets and their support can be computed from the frequent itemsets and their support without any dataset access.

The pseudo-code to determine frequent closed itemsets among frequent itemsets is given in Algorithm 1. Notations are given in Table 7. The input of the algorithm are sets L_i, $1 \leq i \leq k$, containing all frequent itemsets in the dataset. It recursively generates the sets FC_i, $O \leq i \leq k$, of frequent closed i-itemsets from FC_k to FC_0.

Table 7. Notations

L	Set of frequent i-itemsets and their support
FC_i	Set of frequent closed i-itemsets and their support.
isclosed	Variable indicating if the considered itemset is closed or not.

Proposition 1. *The support of a closed itemset is greater than the supports of all its supersets.*

Proof. Let l be a closed i-itemset and s a superset of l. We have $l \subset s \Rightarrow g(l) \supseteq g(s)$ (Property *(1)* of the Galois connection). If $g(l) = g(s)$ then $h(l) = h(s) \Rightarrow l = h(s) \Rightarrow s \subseteq l$ (absurd). It follows that $g(l) \supset g(s) \Rightarrow supp(l) > supp(s)$.

Algorithm 1 Deriving frequent closed itemsets from frequent itemsets.

---

1) $FC_k \leftarrow L_k$;
2) **for** $(i \leftarrow k - 1; i \neq O; i--)$ **do begin**
3) $FC_i \leftarrow \{\}$;
4)   **forall** itemsets $l \in L_i$ **do begin**
5) $isclosed \leftarrow true$;
6)   **forall** itemsets $l' \in L_{i+1}$ **do begin**
7)   **if** $(l \subset l')$ **and** (l.support = l'.support) **then** $isclosed \leftarrow false$;
8)   **end**
9)   **if** ($isclosed = true$) **then** $FC_i \leftarrow FC_i \cup \{l\}$;
10)   **end**
11) **end**
12) $FC_0 \leftarrow \{\emptyset\}$;
13) **forall** itemsets $l \in L_1$ **do begin**
14)   **if** (l.support = $|O|$) **then** $FC_0 \leftarrow \{\}$;
15) **end**

First, the set FC_k is initialized with the set of largest frequent itemsets L_k (step 1). Then, the algorithm iteratively determines which i-itemsets in L_i are closed from L_{k-1} to L_1 (steps 2 to 11). At the beginning of the ith iteration the set FC_i of frequent closed i-itemsets is empty (step 3). In steps 4 to 10, for each frequent itemset l in L_i, we verify that l has the same support as a frequent $(i+1)$-itemset l' in L_{i+1} in which it is included. If so, we have $l' \subseteq h(l)$ and then $l \neq h(l)$: l is not closed (step 7). Otherwise, l is a frequent closed itemset and is inserted in FC_i (step 9). During the last phase, the algorithm determines if the empty itemset is closed by first initializing FC_0 with the empty itemset (step 12) and then considering all frequent 1-itemsets in L_1 (steps 13 to 15). If a 1-itemset l has a support equal to the number of objects in the context, meaning that l is common to all objects, then the itemset $\emptyset$ cannot be closed (we have $supp(\{\emptyset\}) = |O| = supp(l)$) and is removed from FC_0 (step 14). Thus, at the end of the algorithm, each set FC_i contains all frequent closed i-itemsets.

Correctness Since all maximal frequent itemsets are maximal frequent closed itemsets (Lemma 2), the computation of the set FC_k containing the largest frequent closed itemsets is correct. The correctness of the computation of sets FC_i for $i < k$ relies on Proposition 1. This proposition enables one to determine if a frequent i-itemset l is closed by comparing its support and the supports of the frequent $(i+1)$-itemsets in which l is included. If one of them has the same support as l, then l cannot be closed.

4.2. Apriori-Close algorithm

In this section, we present an extension of the Apriori algorithm [AGR 94] computing simultaneously frequent and frequent closed itemsets. The pseudo-code is given in Algorithm 2 and notations in Table 8. The algorithm iteratively generates the sets L_i of frequent i-itemsets from L_1 to L_k. Besides, during the ith iteration, all frequent closed $(i - 1)$-itemsets in FC_{i-1} are determined. The set FC_k is determined during the last step of the algorithm.

Table 8. *Notations*

L_i	Set of frequent i-itemsets, their support and marker *isclosed* indicating if closed or not.
FC_i	Set of frequent closed i-itemsets and their support.

First, the variable k is initialized to 0 (step 1). Then, the set L_1 of frequent 1-itemsets is initialized with the list of items in the context (step 2) and one pass is performed to compute their support (step 3). The set FC_0 is initialized with the empty itemset (step 4) and the supports of itemsets in L_1 are considered (steps 5 to 8). All infrequent 1-itemsets are removed from L_1 (step 6) and if a frequent 1-itemset has a support equal to the number of objects in the context then the empty itemset is removed from FC_0 (step 7). During each of the following iterations (steps 9 to 28), frequent itemsets of size $i+1$, $k > i \geq 1$, and frequent closed itemsets of size i are computed as follows. For all frequent i-itemsets in L_i, the marker *isclosed* is initialized to *true* (step 10). A set L_{i+1} of possible frequent $(i+1)$-itemsets is created by applying the Apriori-Gen function to the set L_i (step 11). For each of these possible frequent $(i+1)$-itemsets, we check that all its subsets of size i exist in L_i (steps 12 to 16). One pass is performed to compute the supports of the remaining itemsets in L_{i+1} (step 17). Then, for each $(i+1)$-itemsets $l \in L_{i+1}$ (steps 18 to 25), if l is infrequent then it is discarded from L_{i+1} (step 19). Otherwise for all i-subsets l' of l, we verify that supports of l' and l are equal; if so, then l' cannot be a closed itemset and its marker *isclosed* is set to false (steps 20 to 24). Then, all frequent i-itemsets in L_i for which marker *isclosed* is true are inserted in the set FC_i of frequent closed i-itemsets (step 26) and the variable k is set to the value of i (step 27). Finally, the set FC_k is initialized with the frequent k-itemsets in L_k (step 29).

Apriori-Gen function The Apriori-Gen function [AGR 94] applies to a set L_i of frequent i-itemsets. It returns a set L_{i+1} of potential frequent $(i+1)$-itemsets. A new itemset in L_{i+1} is created by joining two itemsets in L_i sharing common first i-1 items.

Support-Count function The Support-Count function takes a set L_i of i-itemsets as argument. It efficiently computes the supports of all itemsets $l \in L_i$. Only one dataset pass is required: for each object, o, read, the supports of all itemsets $l \in L_i$ that are included in the set of items associated with o, i.e. $lf(\{o\})$, are incremented. The subsets of $f(\{o\})$ are quickly found using the Subset function described in Section 5.2.

Algorithm 2 Discovering frequent and frequent closed itemsets with Apriori-Close.

---

1) $k \leftarrow 0$;
2) itemsets in $L_1 \leftarrow \{1\text{-itemsets}\}$;
3) $L_1 \leftarrow \text{Support-Count}(L_1)$;
4) $FC_o \leftarrow \{\varnothing\}$;
5) **forall** itemsets $l \in L_1$ **do begin**
6)   **if** (l.support $< minsupp$) **then** $L_1 \leftarrow L_1 \setminus \{l\}$;
7)   **else if** (l.support $= |O|$) **then** $FC_o \leftarrow \{\}$;
8) **end**
9) **for** ($i \leftarrow 1; L_i \neq \{\}; i{+}{+}$) **do begin**
10)   **forall** itemsets $l' \in L_i$ **do** l'.isclosed $\leftarrow$ *true*;
11) $L_{i+1} \leftarrow \text{Apriori-Gen}(L_i)$;
12)   **forall** itemsets $l \in L_{i+1}$ **do begin**
13)   **forall** i-subsets l' of l **do begin**
14)   **if** ($l' \notin L_i$) **then** $L_{i+1} \leftarrow L_{i+1} \setminus \{l\}$;
15)   **end**
16)   **end**
17) $L_{i+1} \leftarrow \text{Support-Count}(L_{i+1})$;
18)   **forall** itemsets $l \in L_{i+1}$ **do begin**
19)   **if** (l.support $< minsupp$) **then** $L_{i+1} \leftarrow L_{i+1} \setminus \{l\}$;
20)   **else do begin**
21)   **forall** i-subsets $l' \in L_i$ of l **do begin**
22)   **if** (l.support $= l'$.support) **then** l'.isclosed $\leftarrow$ *false*;
23)   **end**
24)   **end**
25)   **end**
26) $FC_i \leftarrow \{l \in L_i \mid l.\text{isclosed} = true\}$;
27) $k \leftarrow i$;
28) **end**
29) $FC_k \leftarrow L_k$;

Correctness Since the support of a frequent closed itemset l is different from the support of all its supersets (Proposition 1), the computation of sets FC_i for $i < k$ is correct. Hence, a frequent i-itemset $l' \in L_i$ is determined closed or not by comparing its support with the supports of all frequent $(i + 1)$-itemsets $l \in L_{i+1}$ for which $l' \subset l$. Lemma 2 ensures the correctness of the computation of the set FC_k containing the largest frequent closed itemsets.

Example 7. Figure 2 illustrates the execution of the Apriori-Close algorithm with the context $\mathcal{D}$ for a minimum support of 2/5.

5. Generating bases for association rules

In Section 5.1, we present an algorithm to generate the Duquenne-Guigues basis for exact association rules. In Sections 5.2 and 5.3 algorithms are described

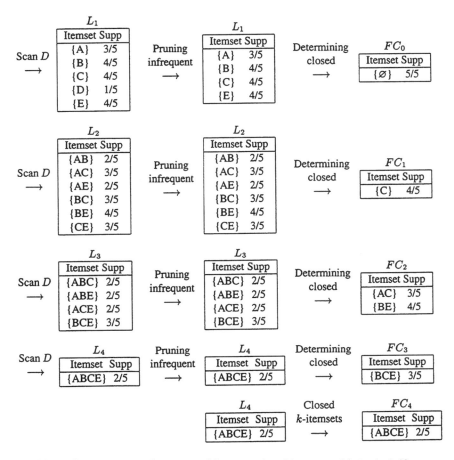

Figure 2. *Discovering frequent and frequent closed itemsets with Apriori-Close*

achieving the proper basis and the structural basis for approximate association rules respectively.

5.1. Generating Duquenne-Guigues basis for exact association rules

The pseudo-code generating the Duquenne-Guigues basis for exact association rules is given in Algorithm 3. Notations are given in Table 9. The algorithm takes as input the sets L_i, $1 \le i \le k$, containing the frequent itemsets and their support, and the sets FC_i, $0 \le i \le k$, containing the frequent closed itemsets and their support. It first computes the frequent pseudo-closed itemsets iteratively (steps 2 to 17) and then uses them to generate the Duquenne-Guigues basis for exact association rules DG (steps 18 to 22).

Table 9. *Notations*

L_i	Set of frequent i-itemsets and their support.
FC_i	Set of frequent closed i-itemsets and their support.
FP_i	Set of frequent pseudo-closed i-itemsets, their closure and their support.
DG	Duquenne-Guigues basis for exact association rules.

First, the set DG is initialized to the empty set (step 1). If the empty itemset is not a closed itemset (it is then necessarily a pseudo-closed itemset), it is inserted in FP_0 (step 2). Otherwise FP_0 is empty (step 3). Then, the algorithm recursively determines which i-itemsets in L_i are pseudo-closed from L_1 to L_k (steps 4 to 16). At each iteration, the set FP_i is initialized with the list of frequent i-itemsets that are not closed (step 5) and each frequent i-itemset l in FP_i is considered as follows (steps 6 to 15). The variable *pseudo* is set to *true* (step 7). We verify for each frequent pseudo-closed itemset p previously discovered (i.e. in FP_j with $j < i$) if p is contained in l (steps 8 to 13). In that case and if the closure of p is not included in l, then l is not pseudo-closed and is removed from FP_i (steps 9 to 12). Otherwise, the closure of l (i.e. the smallest frequent closed itemset containing l) is determined (step 14). Once all frequent pseudo-closed itemsets p and their closure are computed, all rules with the form $r: p \Rightarrow (p.\text{closure} \setminus p)$ are generated (steps 17 to 21). The algorithm results in the set DG containing all rules in the Duquenne-Guigues basis for exact association rules.

Algorithm 3 Generating Duquenne-Guigues basis for exact association rules.

```
 1) DG ← {}
 2) if (FC₀ = {}) then FP₀ ← {∅};
 3) else FP₀ ← {};
 4) for (i ← 1; i ≤ k; i++) do begin
 5)    FPᵢ ← Lᵢ \ FCᵢ;
 6)    forall itemsets l ∈ FPᵢ do begin
 7)       pseudo ← true;
 8)       forall itemsets p ∈ FPⱼ with j < i do begin
 9)          if (p ⊂ l) and (p. closure ⊄ l) then do begin
10)             pseudo ← false;
11)             FPᵢ ← FPᵢ \ {l};
12)          end
13)       end
14)       if (pseudo = true) then l.closure ← Min⊆ ({c ∈ FCⱼ >ᵢ | l ⊆ c});
15)    end
16) end
17) forall sets FPᵢ where FPᵢ ≠ {} do begin
18)    forall pseudo-closed itemsets p ∈ FPᵢ do begin
19)       DG ← DG ∪ {r: p ⇒ (p.closure\p),p.support};
20)    end
21) end
```

Correctness Since the itemset $\varnothing$ has no subset, if it is not a closed itemset then it is by definition a pseudo-closed itemset and the computation of the set FP_0 is correct. The correctness of the computation of frequent pseudo-closed i-itemsets in FP_i for $1 \leq i \leq k$ relies on Definition 7. All frequent i-itemsets l in L_i that are not closed, i.e. not in FC_i, are considered. Those l containing the closures of all frequent pseudo-closed itemsets that are subsets of l are inserted in FP_i. According to Definition 7, these i-itemsets are all frequent pseudo-closed i-itemsets and the sets FP_i are correct. The association rules generated in the last phase of the algorithm are all rules with a frequent pseudo-closed itemset in the antecedent. Then, the resulting set DG corresponds to the rules in the Duquenne-Guigues basis for exact association rules characterized in Definition 8.

Example 8. Figure 3 shows the generation of the Duquenne-Guigues basis for exact association rules from the context $\mathcal{D}$ for a minimum support of 2/5.

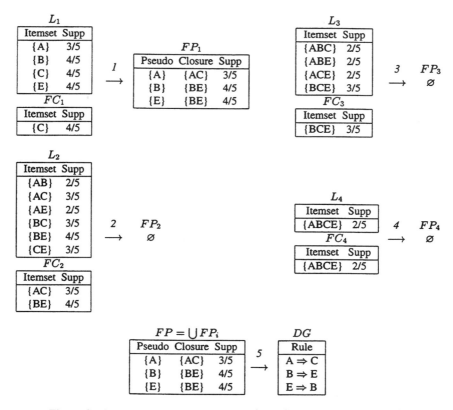

Figure 3. *Generating Duquenne-Guigues basis for exact association rules*

5.2. Generating proper basis for approximate association rules

The pseudo-code generating the proper basis for approximate association rules is presented in Algorithm 4. Notations are given in Table 10. The algorithm takes as input the sets FC_i, $1 \leq i \leq k$, containing the frequent closed non-empty itemsets and their support. The output of the algorithm is the proper basis for approximate association rules PB.

Table 10. Notations

FC_i	Set of frequent closed i-itemsets and their support.
S_j	Set of j-itemsets that are subsets of the considered itemset.
PB	Proper basis for approximate association rules.

The set PB is first initialized to the empty set (step 1). Then, the algorithm iteratively considers all frequent closed itemsets $l \in FC_i$ for $2 \leq i \leq k$. It determines which frequent closed itemsets $l' \in FC_{j<i}$ are subsets of l and generates association rules with the form $l' \to l\backslash l'$ that have sufficient confidence (steps 2 to 12) as follows. During the ith iteration, each itemset l in FC_i is considered (steps 3 to 11). For each set FC_j, $1 \leq j < i$, a set S_j containing all frequent closed j-itemsets in FC_j that are subsets of l is created (step 5). Then, for each of these subsets $l' \in S_j$ (steps 6 to 9), we compute the confidence of the proper approximate association rule r: $l' \to l\backslash l'$ (step 7). If the confidence of r is sufficient then r is inserted in PB (steps 8 to 9). At the end of the algorithm, the set PB contains all rules of the proper basis for approximate association rules.

Subset function The subset function takes a set X of itemsets and an itemset y as arguments. It determines all itemsets $x \in X$ that are subsets of y. In algorithm implementation, frequent and frequent closed itemsets are stored in a *prefix-tree* structure [PAS 98, PAS 99a] in order to improve efficiency of the subset search.

Algorithm 4 Generating proper basis for approximate association rules.

```
 1) PB ← {}
 2) for (i ← 2; i ≤ k; i++) do begin
 3)    forall itemsets l ∈ FCi do begin
 4)       for (j ← i – 1; j > 0; j– –) do begin
 5)          Sj ← Subsets (FCj, l);
 6)          forall itemsets l' ∈ Sj do begin
 7)             conf(r) ← l.support /l'.support;
 8)             if (conf(r) ≥ minconf)
 9)             then PB ← PB ∪ {r: l' → l\l', l.support, conf(r)};
10)          end
11)       end
12)    end
13) end
```

Correctness The correctness of the algorithm relies on the fact that we inspect all proper approximate association rules holding in the dataset. For each frequent closed itemset, the algorithm computes, among its subsets, all other frequent closed itemsets. Then, the generation of all rules between two frequent closed itemsets having sufficient confidence is ensured. These rules are all proper approximate association rules holding in the dataset, and the resulting set *PB* is the proper basis for approximate association rules defined in Theorem 3.

Example 9. Figure 4 shows the generation of the proper basis for approximate association rules in the context $\mathcal{D}$ for a minimum support of 2/5 and a minimum confidence of 1/2.

5.3. Generating structural basis for approximate association rules

The pseudo-code generating the structural basis for approximate association rules is given in Algorithm 5. Notations are given in Table 11. The algorithm takes as

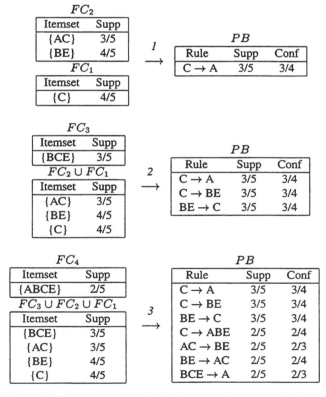

Figure 4. Generating proper basis for approximate association rules

---

Algorithm 5   Generating structural basis for approximate association rules.

1) $SB \leftarrow \{\}$;
2) **for** $(i \leftarrow 2; i \leq k; i++)$ **do begin**
3)   **forall** itemsets $l \in FC_i$ **do begin**
4) $CR \leftarrow \{\}$;
5)   **for** $(j \leftarrow i-1; j > 0; j--)$ **do begin**
6) $S_j \leftarrow$ Subsets (FC_j, l);
7)   **end**
8)   **for** $(j \leftarrow i-1; j > 0; j--)$ **do begin**
9)   **forall** itemsets $l' \in S_j$ **do begin**
10) $conf(r) \leftarrow l.support/l'.support$;
11)   **if** $(conf(r) \geq minconf)$ **then do begin**
12) $CR \leftarrow CR \cup \{r: l' \rightarrow l\backslash l', l.support, conf(r)\}$;
13)   **for** $(n \leftarrow j-1; n > 0; n--)$ **do begin**
14) $S_n \leftarrow S_n\backslash Subsets(S_n, l')$;
15)   **end**
16)   **end**
17)   **end**
18)   **end**
19)   **if** $(CR \neq \{\})$ **then do begin**
20) $maxconf \leftarrow Max_{r \in CR}(conf(r))$;
21)   **find first** $\{r \in CR \mid conf(r) = maxconf\}$;
22) $SB \leftarrow SB \cup \{r\}$;
23)   **end**
24)   **end**
25) **end**

---

input the sets FC_i, $1 \leq i \leq k$, of frequent closed non-empty itemsets and their support. It generates the structural basis for approximate association rules SB represented by the maximal confidence spanning forest $\mathcal{F}_{FC}$ associated with $FC = \cup_{i=1}^{i=k} FC_i$ (without the empty itemset).

Table 11. *Notations*

FC_i	Set of frequent closed i-itemsets and their support.
S_j	Set of j-itemsets that are subsets of the itemset considered.
CR	Set of candidate approximate association rules.
SB	Structural basis for approximate association rules.

The set SB is first initialized to the empty set (step 1). Then, the algorithm iteratively considers all frequent closed itemsets $l \in FC_i$ for $2 \leq i \leq k$. It determines which frequent closed itemsets $l' \in FC_{j<i}$ are covered by l, i.e. are direct predecessors of l, and then generates the maximal confidence association rules with the form $l \rightarrow l' \backslash l$ that hold (steps 2 to 25). During the ith iteration, each itemset l in FC_i is considered (steps 3 to 24) as follows. The set CR of candidate association rules with l in the consequent is initialized to the empty set (step 4). For $1 \leq j < i$, sets S_j containing all frequent closed j-itemsets in FC_j that are subsets of l are created (steps 5 to 7). Then, all these subsets of l are considered in decreasing order of their size (steps 8 to 18). For each of these subsets $l' \in S_j$, the confidence of the proper approximate association rule r: $l' \to l \setminus l'$ is computed (step 10). If the confidence of r is sufficient, r is inserted in CR (step 12) and all subsets l'' of l' are removed from $S_{n<j}$ (steps 13 to 15). This is because rules with the form $l'' \to \Lambda l''$ with $l'' \in S_{n<j}$ are transitive proper approximate rules. Finally, the candidate proper approximate rules with l in the consequent that are in CR are pruned (steps 19 to 23): the maximum confidence value *maxconf* of rules in CR is determined (step 20) and the first rule with such a confidence is inserted in SB (steps 21 and 22). At the end of the algorithm, the set SB thus contains all rules in the structural basis for approximate association rules.

Correctness The algorithm considers all association rules $l' \to \Lambda l'$ with confidence $\geq$ *minconf* between two frequent closed itemsets l and l' where l covers l'. These rules are all proper non-transitive approximate association rules that hold and can be represented by the edges of the graph G_{FC} (Definition 8) without transitive edges. Moreover, among all rules with the form $X \to \Lambda X$ (generated from l), we keep only the first one with confidence equal to the maximal confidence of rules $X \to \Lambda X$. Only preserving this rule is equivalent to the cycle removing in the graph G_{FC} in the same manner as explained in Definition 9. Then, the resulting set SB can be represented as the maximal confidence spanning forest $\mathcal{F}_{FC}$ without edges from the empty itemset. SB contains all rules in the structural basis for approximate association rules defined in Theorem 4.

Example 10. Figure 5 depicts the generation of the structural basis for approximate association rules in the context $\mathcal{D}$ for a minimum support of 2/5 and a minimum confidence of 1/2.

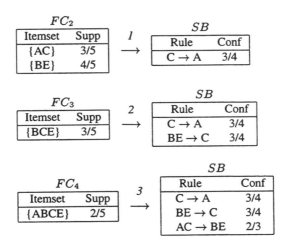

Figure 5. *Generating structural basis for approximate association rules*

6. Experimental results

Experiments were performed on a Pentium II PC with a 350 Mhz clock rate, 128 MBytes of RAM, running the Linux operating system. Algorithms were implemented in C++. Characteristics of the datasets used are given in Table 12. These datasets are the T10I4D100K[4] synthetic dataset that mimics market basket data, the C20D10K and the C73D10K census datasets from the PUMS sample file[5], and the MUSHROOMS[6] dataset describing mushroom characteristics. In all experiments, we attempted to choose significant minimum support and confidence threshold values: we observed threshold values used in other papers for experiments on similar data types and inspected rules extracted in the bases.

Table 12. *Datasets*

Name	Number of objects	Average size of objects	Number of items
T10I4D100K	100,000	10	1,000
MUSHROOMS	8,416	23	127
C20D10K	10,000	20	386
C73D10K	10,000	73	2,177

6.1. Relative performance of Apriori and Apriori-Close

We conducted experiments to compare response times obtained with Apriori and Apriori-Close on the four datasets. Results for the T10I4D100K and MUSHROOMS datasets are presented in Table 13. We can observe that execution times are identical for the two algorithms: adding the frequent closed itemset derivation to the frequent itemset discovery does not induce additional computation time. Similar results were obtained for C20D10K and C73D10K datasets.

Table 13. *Execution times of Apriori and Apriori-Close*

Minsupp	Apriori	Apriori-Close	Minsupp	Apriori	Apriori-Close
2.0%	1.99s	1.97s	90%	0.28s	0.28s
1.0%	3.47s	3.46s	70%	0.73s	0.73s
0.5%	9.62s	9.70s	50%	2.40s	2.70s
0.25%	15.02s	14.92s	30%	18.22s	17.93s

T10I4D100K	MUSHROOMS

6.2. Number of rules and execution times of the rule generation

Table 14 shows the total number of exact association rules and their number in the Duquenne-Guigues basis for exact rules. Table 15 shows the total number of approximate association rules, their number in the proper basis and in the struc-

---

[4] http://www.almaden.ibm.com/cs/quest/syndata.html
[5] ftp://ftp2.cc.ukans.edu/pub/ippbr/census/pums/pums90ks.zip
[6] ftp://ftp.ics.uci.edu/~cmerz/mldb.tar.Z

Table 14. Number of exact association rules extracted

Dataset	Minsupp	Exact rules	Duquenne-Guigues basis
T1014D100K	0.5%	0	0
MUSHROOMS	30%	7,476	69
C20D10K	50%	2,277	11
C73D10K	90%	52,035	15

Table 15. Number of approximate association rules extracted

Dataset (Minsupp)	Minconf rules	Approximate basis	Proper basis	Non-transitive basis	Structural
	90%	16,260	16,260	3,511	916
T10I4D100K	70%	20,419	20,419	4,004	1,058
(0.5%)	50%	21,686	21,686	4,191	1,140
	30%	22,952	22,952	4,519	1,367
	90%	12,911	806	563	313
MUSHROOMS	70%	37,671	2,454	968	384
(30%)	50%	56,703	3,870	1,169	410
	30%	71,412	5,727	1,260	424
	90%	36,012	4,008	1,379	443
C20D10K	70%	89,601	10,005	1,948	455
(50%)	50%	116,791	13,179	1,948	455
	30%	116,791	13,179	1,948	455
	95%	1,606,726	23,084	4,052	939
C73D10K	90%	2,053,896	32,644	4,089	941
(90%)	85%	2,053,936	32,646	4,089	941
	80%	2,053,936	32,646	4,089	941

tural basis for approximate rules, and the number of non-transitive rules in the proper basis for approximate rules (5th column). For example in the context $\mathcal{D}$, rules $C \rightarrow A$ and $AC \rightarrow BE$ are extracted, as well as the rule $C \rightarrow ABE$ which is clearly transitive. Since by construction, its confidence – retrieved by multiplying the confidences of the two former – is less than theirs, this rule is the less interesting among the three. Reducing the extraction to non-transitive rules in the proper basis for approximate rules can also be interesting. Such rules are generated by a variant of Algorithm 5 with the last pruning strategy (steps 20 and 21) removed: all candidate rules in CR are inserted in SB.

Table 16 shows for the four datasets the average relative size of bases compared with the sets of all rules obtained. In the case of weakly correlated data (T10I4D100K), no exact rule is generated and the proper basis for approximate rules contains all approximate rules that hold. The reason is that, in such data, all frequent itemsets are frequent closed itemsets. In the case of correlated data (MUSHROOMS, C20D10K and C73D10K), the number of extracted rules in bases is much smaller than the total number of rules that hold.

Table 16. *Average relative size of bases*

Dataset	Duquenne-Guigues basis	Proper basis	Non-transitive basis	Structural basis
T10I4D100K	–	100.00%	20.05%	5.49%
MUSHROOMS	0.92%	6.90%	2.69%	1.19%
C20D10K	0.48%	11.21%	2.33%	0.63%
C73D10K	0.03%	1.55%	0.21%	0.05%

Figure 6 shows for each dataset the execution times of the computation of all rules (using the algorithm described in [AGR 94]) and bases. Execution times of the derivation of the Duquenne-Guigues basis for exact rules and the proper basis for non-transitive approximate rules are not presented since they are identical to those of the derivation of the Duquenne-Guigues basis for exact rules and the structural basis for approximate rules (*Duquenne-Guigues and structural bases*).

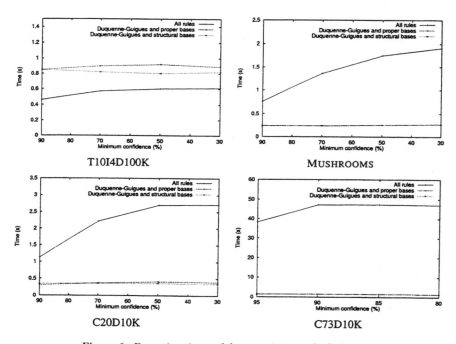

Figure 6. *Execution times of the association rule derivation*

7. Conclusions

In this paper, we present new algorithms for efficiently generating bases for association rules. A basis is a set of non-redundant rules from which all association rules can be derived; thus it captures all useful information. Moreover, its size is significantly reduced compared with the set of all possible rules because redundant, and thus useless, rules are discarded. Our approach has a twofold advantage: on one hand, the user is provided with a smaller set of resulting rules, easier to handle, and vehiculing information of improved quality. On the other hand, execution times are reduced compared with the discovery of all association rules. Such results are proved (in the groundwork of lattice theory and data analysis) and illustrated by experiment, achieved from real-life datasets.

Integrating reduction methods Templates, as defined in [BAR 97, KLE 94], can directly be used for extracting from the bases all association rules matching some user-specified patterns. Information in taxonomies associated with the dataset can also be integrated in the process as proposed in [HAN 95, SRI 95] for extracting bases for generalized (multi-level) association rules. Integrating item constraints and statistical measures, such as described in [BAY 99b, NG 98, SRI 97] and [BRI 97a, PIA 91] respectively, in the generation of bases, requires further work.

Functional and approximate dependencies Algorithms presented in this paper can be adapted to generate bases for functional and approximate dependencies. In [HUH 98, LOP 00, MAN 94], such bases and algorithms for generating them were proposed. However, the Duquenne-Guigues basis is smaller than the basis for functional dependencies constituted of minimal non-trivial functional dependencies. Hence, the number of rules in the Duquenne-Guigues basis is minimal [DEM 92, GAN 99]. Furthermore, the proper and structural bases for approximate rules are also smaller than the basis for approximate dependencies defined in [HUH 98]. Adapting our algorithms to the discovery of functional and approximate dependencies is an ongoing research topic.

Minimal non-redundant association rules The bases for association rules defined in this paper significantly reduce the number of extracted rules and give a high quality non-redundant summary of valid association rules. However, they are not constituted of the non-redundant rules with minimal antecedent and maximal consequent, called *minimal non-redundant association rules*. Such rules are the most informative, since they provide a minimal set that maximizes the information conveyed, and can be characterized using frequent closed itemsets and their generators. This is demonstrated in [BAS 00, PAS 00] and algorithms for generating them using frequent closed itemsets and their generators, such as extracted by the Close or the A-Close algorithm, or using the frequent itemsets, for extending an existing implementation, are proposed. Results of experiments conducted on real-life datasets are exhibited and show that this generation is both efficient and useful.

Acknowledgements

The authors would like to gratefully acknowledge Rosine Cicchetti and Mohand-Saïd Hacid for their constructive comments.

REFERENCES

[AGR 93] ACRAWAL R., IMIELINSKI T., SWAMI A., "Mining Association Rules Between Sets of Items in Large Databases", *Proc. of the ACM SIGMOD Conference*, Washington, May 26–28, 1993, p. 207–216.

[AGR 94] AGRAWAL R., SRIKANT R., "Fast Algorithms for Mining Association Rules", *Proc. of the 20th VLDB Conference*, Santiago de Chile, September 12–15, 1994, p. 478–499, Expanded version in IBM Research Report RJ9839.

[BAR 97] BARALIS E., PSAILA G., "Designing Templates for Mining Association Rules", *Journal of Intelligent Information Systems*, vol. 9, num. 1, 1997, p. 7–32, Kluwer Academic Publishers.

[BAS 00] BASTIDE Y., PASQUIER N., TAOUIL R., STUMME G., LAKHAL L., "Mining Minimal Non-redundant Association Rules Using Frequent Closed Itemsets", *Proc. of the 6th DOOD Conference*, London, July 24–28, 2000.

[BAY 98] BAYARDO R.J., "Efficiently Mining Long Patterns from Databases", *Proc. of the ACM SIGMOD Conference*, Washington, June 2–4, 1998, p. 85–93.

[BAY 99a] BAYARDO R.J., AGRAWAL R., "Mining the Most Interesting Rules", *Proc. of the 5th KDD Conference*, San Diego, August 15–18, 1999, p. 145–154.

[BAY 99b] BAYARDO R.J., AGRAWAL R., GUNOPULOS D., "Constraint-Based Rule Mining in Large, Dense Databases", *Proc. of the 15th ICDE Conference*, Sydney, March 23–26, 1999, p. 188–197.

[BIR 67] BIRKHOFF G., "Lattices Theory", *Colloquium Publications XXV*, American Mathematical Society, 1967, Third edition.

[BRI 97a] BRIN S., MOTWANI R., SILVERSTEIN C., "Beyond Market Baskets: Generalizing Association Rules to Correlation", *Proc. of the ACM SIGMOD Conference*, Tucson, May 13–15, 1997, p. 265–276.

[BRI 97b] BRIN S., MOTWANI R., ULLMAN J.D., TSUR S., "Dynamic Itemset Counting and Implication Rules for Market Basket Data", *Proc. of the ACM SIGMOD Conference*, Tucson, May 13–15, 1997, p. 255–264.

[BUR 98] BURMEISTER P., "Formal Concept Analysis with CONIMP: Introduction to the Basic Features", report, 1998, Technische Hochschule Darmstadt, Germany.

[DEM 92] DEMETROVICS J., LIBKIN L., MUCHNIK I.B., "Functional Dependencies in Relational Databases: A Lattice Point of View", *Discrete Applied Mathematics*, vol. 40, 1992, p. 155–185.

[DUQ 86] DUQUENNE V., GUIGUES J.-L., "Famille Minimale d'Implication Informatives Résultant d'un Tableau de Données Binaires", *Mathématiques et Sciences Humaines*, vol. 24, num. 95, 1986, p. 5–18.

[GAN 91] GANTER B., REUTER K., "Finding all Closed Sets: A General Approach", *Order*, 1991, p. 283–290, Kluwer Academic Publishers.

[GAN 99] GANTER B., WILLE R., *Formal Concept Analysis: Mathematical Foundations*, Springer, 1999.

[HAN 95] HAN J., FU Y., "Discovery of Multiple-Level Association Rules from Large Databases", *Proc. of the 21st VLDB Conference*, Zurich, September 11–15, 1995, p. 420–431.

[HEC 96] HECKERMAN D., "Bayesian Networks for Knowledge Discovery", *Advances in Knowledge Discovery and Data Mining*, 1996, p. 273–305, AAAI Press.

[HUH 98] HUHTALA Y., KÄRKKÄINEN J., PORKKA P., TOIVONEN H., "Efficient Discovery of Functional and Approximate Dependencies Using Partitions", *Proc. of the 14th ICDE Conference*, Orlando, February 23–27, 1998, p. 392–401.

[KLE 94] KLEMETTINEN M., MANNILA H., RONKAINEN P., TOIVONEN H., VERKAMO A.I., "Finding Interesting Rules from Large Sets of Discovered Association Rules", *Proc. of the 3rd CIKM Conference*, Gaithersburg, November 29–December 2, 1994, p. 401–407.

[LIN 98] LIN D., KEDEM Z.M., "Pincer-Search: A New Algorithm for Discovering the Maximum Frequent Set", *Proc. of the 6th EDBT Conference*, Valencia, March 23–27, 1998, p. 105–119.

[LIU 97] LIU B., HSU W., CHEN S., "Using General Impressions to Analyse Discovered Classification Rules", *Proc. of the 3rd KDD Conference*, Newport Beach, August 14–17, 1997, p. 31–36.

[LOP 00] LOPES S., PETIT J.-M., LAKHAL L., "Efficient Discovery of Functional Dependencies and Armstrong Relations", *Proc. of the 7th EDBT Conference*, Konstanz, March 27–31, 2000.

[LUX 91] LUXENBURGER M., "Implications Partielles dans un Contexte", *Mathématiques, Informatique et Sciences Humaines*, vol. 29, num. 113, 1991, p. 35–55.

[MAN 94] MANNILA H., RÄIMA K.J., "Algorithms for Inferring Functional Dependencies from Relations", *Data & Knowledge Engineering*, vol. 12, num. 1, 1994, p. 83–99.

[MEO 96] MEO R., PSAILA G., CERI S., "A New SQL-Like Operator for Mining Association Rules", *Proc. of the 22nd VLDB Conference*, Bombay, September 3–6, 1996, p. 122–133.

[MOR 98] MORIMOTO Y., FUKUDA T., MATSUZAWA H., TOKUYAMA T., YODA K., "Algorithms for Mining Association Rules for Binary Segmentations of Huge Categorical Databases", *Proc. of the 24th VLDB Conference*, New York City, August 24–27, 1998, p. 380–391.

[NG 98] NG R. T., LAKSHMANAN V. S., HAN J., PANG A., "Exploratory Mining and Pruning Optimizations of Constrained Association Rules", *Proc. of the ACM SIGMOD Conference*, Washington, June 2–4, 1998, p. 13–24.

[PAS 98] PASQUIER N., BASTIDE Y., TAOUIL R., LAKHAL L., "Pruning Closed Itemset Lattices for Association Rules", *Proc. of the 14th BDA Conference*, Hammamet, October 26–30, 1998, p. 177–196.

[PAS 99a] Pasquier N., Bastide Y., Taouil R., Lakhal L., "Efficient Mining of Association Rules using Closed Itemset Lattices", *Information Systems*, vol. 24, num. 1, 1999, p. 25–46, Elsevier Science.

[PAS 99b] Pasquier N., Bastide Y., Taouil R., Lakhal L., "Discovering Frequent Closed Itemsets for Association Rules", *Proc. of the 7th ICDT Conference*, Jerusalem, January 10–12, 1999, p. 398–416.

[PAS 00] Pasquier N., "Mining Association Rules Using Formal Concept Analysis", *Proc. of the 8th ICCS Conference*, Darmstadt, August 14–18, 2000.

[PIA 91] Piatetsky-Shapiro G., "Discovery, Analysis, and Presentation of Strong Rules", *Knowledge Discovery in Databases*, 1991, p. 229–248, AAAI Press.

[SAV 95] Savasere A., Omiecinski E., Navathe S., "An Efficient Algorithm for Mining Association Rules in Large Databases", *Proc. of the 21st VLDB Conference*, Zurich, September 11–15, 1995, p. 432–444.

[SIL 96] Silberschatz A., Tuzgilin A., "What Makes Patterns Interesting in Knowledge Discovery Systems", *IEEE Transactions on Knowledge and Data Engineering*, vol 8, num. 6, 1996, p. 970–974.

[SIL 98] Silverstein C., Brin S., Motwani R., "Beyond Market Baskets: Generalizing Association Rules to Dependence Rules", *Data Mining and Knowledge Discovery*, vol. 2, num. 1, 1998, p. 39–68.

[SRI 95] Srikant R., Agrawal R., "Mining Generalized Association Rules", *Proc. of the 21st VLDB Conference*, Zurich, September 11–15, 1995, p. 407–419.

[SRI 96] Srikant R., Agrawal R., "Mining Quantitative Association Rules in Large Relational Tables", *Proc. of the ACM SIGMOD Conference*, Montreal, June 4–6, 1996, p. 1–12.

[SRI 97] Srikant R., Agrawal R., "Mining Association Rules with Item Constraints", *Proc. of the 3rd KDD Conference*, Newport Beach, August 14–17, 1997, p. 67–73.

[TOI 95] Toivonen H., Klemettinen M., Ronkainen P., Hätönen K., Mannila H., "Pruning and Grouping of Discovered Association Rules", *Notes of the ECML'95 Workshop*, Heraklion, April 1995, p. 47–52.

[TOI 96] Toivonen H., "Sampling Large Databases for Association Rules", *Proc. of the 22nd VLDB Conference*, Bombay, September 3–6, 1996, p. 134–145.

[WIL 92] Wille R., "Concept Lattices and Conceptual Knowledge Systems", *Computers and Mathematics with Applications*, vol. 23, 1992, p. 493–515.

[ZAK 97] Zaki M. J., Parthasarathy S., Ogihara M., Li W., "New Algorithms for Fast Discovery of Association Rules", *Proc. of the 3rd KDD Conference*, Newport Beach, August 14–17, 1997, p. 283–286.

Index

Innovative Technology Series
Information Systems and Networks

Other titles in this series

Advances in UMTS Technology

Edited by J. C. Bic and E. Bonek
£58.00   1903996147   216 pages   April 2002

The Universal Mobile Telecommunication System (UMTS), the third generation mobile system, is now coming into use in Japan and Europe. The main benefits – spectrum efficient radio interfaces offering high capacity, large bandwidths, ability to interconnect with IP-based networks, and flexibility of mixed services with variable data – offer exciting prospects for the deployment of these networks.

This publication, written by academic researchers, manufacturers and operators, addresses several issues emphasising future evolution to improve the performance of the 3rd generation wireless mobile on to the 4th generation. Outlining as it does key topics in this area of enormous innovation and commercial investment, this material is certain to excite considerable interest in academia and the communications industry.

The content of this book is derived from *Annals of Telecommunications*, published by GET, Direction Scientifique, 46 rue Barrault, F 75634 Paris Cedex 13, France.

Java and Databases

Edited by A. Chaudhri
£35.00   1903996155   136 pages   April 2002

Many modern data applications such as geographical information systems, search engines and computer aided design systems depend on having adequate storage management control. The tools required for this are called persistent storage managers. This book describes the use of the programming language Java in these and other applications.

This publication is based on material presented at a workshop entitled 'Java and Databases: Persistence Options' held in Denver, Colorado in November 1999. The contributions represent the experience acquired by academics, users and practitioners in managing persistent Java objects in their organisations.

For information about other engineering and science titles published by Hermes Penton Science, go to **www.hermespenton.com**

Quantitative Approaches in Object-oriented Software Engineering

Edited by F. Brito e Abreu, G. Poels, H. Sahraoui, H. Zuse
£35.00   1903996279   136 pages   April 2002

Software internal attributes have been extensively used to help software managers, customers and users characterise, assess and improve the quality of software products. Software measures have been adopted to increase understanding of how software internal attributes affect overall software quality, and estimation models based on software measures have been used successfully to perform risk analysis and to assess software maintainability, reusability and reliability. The object-oriented approach presents an advance in technology, providing more powerful design mechanisms and new technologies including OO frameworks, analysis/design patterns, architectures and components. All have been proposed to improve software engineering productivity and software quality.

The key topics in this publication cover metrics collection, quality assessment, metrics validation and process management. The contributors are from leading research establishments in Europe, South America and Canada.

Turbo Codes: Error-correcting Codes of Widening Application

Edited by M. Jézéquel and R. Pyndiah
£50.00   1903996260   206 pages   May 2002

The last ten years have seen the appearance of a new type of correction code – the *turbo code*. This represents a significant development in the field of error-correcting codes.

The decoding principle is to be found in an iterative exchange of information (*extrinsic information*) between elementary decoders. The turbo concept is now applied to block codes as well as other parts of a digital transmission system, such as detection, demodulation and equalisation.

Providing an excellent compromise between complexity and performance, turbo codes have now become a reference in the field, and their range of application is increasing rapidly to mobile communications, interactive television, as well as wireless networks and local radio loops. Future applications could include cable transmission, short distance communication or data storage.

This publication includes contributions from an internationally-based group of authors, from France, Sweden, Australia, USA, Italy, Germany and Norway.

The content of this book is derived from *Annals of Telecommunications*, published by GET, Direction Scientifique, 46 rue Barrault, F 75634 Paris Cedex 13, France.

For information about other engineering and science titles published by Hermes Penton Science, go to **www.hermespenton.com**

Millimeter Waves in Communication Systems

Edited by M. Ney
£50.00   1903996171   180 pages   May 2002

The topics covered in this publication provide a summary of major activities in the development of components, devices and systems in the millimetre-wave range. It shows that solutions have been found for technological processes and design tools needed in the creation of new components. Such developments come in the wake of the demands arising from frequency allocations in this range. The other numerous new applications include satellite communication and local area networks that are able to cope with the ever-increasing demand for faster systems in the telecommunications area.

The content of this book is derived from *Annals of Telecommunications*, published by GET, Direction Scientifique, 46 rue Barrault, F 75634 Paris Cedex 13, France.

Intelligent Agents for Telecommunication Environments

Edited by D. Gaïti and O. Martikainen
£35.00   1903996295   110 pages   June 2002

Telecommunication systems become more dynamic and complex with the introduction of new services, mobility and active networks. The use of artificial intelligence and intelligent agents, integrated reasoning, learning, co-operating and mobility capabilities to provide predictive control are among possible ways forward. There is a need to investigate performance, flow and congestion control, intelligent control environment, security service creation and deployment and mobility of users, terminals and services. New approaches include the introduction of intelligence in nodes and terminal equipment in order to manage and control the protocols, and the introduction of intelligence mobility in the global network. These tools aim to provide the quality of service and adapt the existing infrastructure to be able to handle the new functions and achieve the necessary co-operation between nodes. This book's contributors, who come from research establishments all over the world, address these problems and provide ways forward in this fast-developing area of intelligence in networks.

For information about other engineering and science titles published by Hermes Penton Science, go to **www.hermespenton.com**

Video Data

Edited by M-S Hacid and S. Hassas
£35.00   1903996228   128 pages   July 2002

With recent progress in computer technology and reduction in processing costs it is possible to store huge amounts of video data needed in today's communication applications. To obtain efficient use of such data efficient storage, querying and navigation of this data is needed. To meet the increasing demands of the new developments, new management techniques and tools need to be developed, and this publication addresses the application of the many research disciplines involved.

Applications and Services in Wireless Networks

Edited by H. Afifi and D. Zeghlache
£58.00   1903996309   260 pages   July 2002

Emerging wireless technologies for both public and private use have led to the creation of new applications. These include the adaptation of current network management procedures and protocols and the introduction of unified open service architectures. Aspects such as accounting for multiple media access and QoS (Quality of Service) profiling must also be introduced to enable multimedia service offers, service management and service control over the wireless Internet. Security and content production are needed to foster the development of new services while adaptable applications for variable bandwidth and variable costs will open new possibilities for ubiquitous communications. In this book the contributors, drawn from a broad international field, address these prospects from the most recent perspectives.

For information about other engineering and science titles published by Hermes Penton Science, go to **www.hermespenton.com**

Mobile Agents for Telecommunication Applications

Edited by E. Horlait
£35.00   1903996287   110 pages   July 2002

Mobile agents are concerned with self-contained and identifiable computer programs that can move within a network and can act on behalf of the user and another entity. Most current research work on the mobile agent paradigm has two general goals: the reduction of network traffic and asynchronous interaction, the object being to reduce information overload and to efficiently use network resources. The international contributors to this book provide an overview of how the mobile code can be used in networking with the aim of developing further intelligent information retrieval, network and mobility management, and network services.

Wireless Mobile Phone Access to the Internet

Edited by Thomas Noel
£40.00   1903996325   150 pages   August 2002

Wireless mobile phone access to the Internet will add a new dimension to the way we access information and communicate. This book is devoted to the presentation of recent research on the deployment of the network protocols and services for mobile hosts and wireless communication on the Internet.

A lot of wireless technologies have already appeared: IEEE 802.11b, Bluetooth, HiperLAN/2, GPRS, UTMS. All of them have the same goal: offering wireless connectivity with minimum service disruption between mobile handovers. The mobile world is divided into two parts: firstly, mobile nodes can be attached to several access points when mobiles move around; secondly, ad-hoc networks exist which do not use any infrastructure to communicate. With this model all nodes are mobiles and they cooperate to forward information between each other. This book presents these two methods of Internet access and presents research papers that propose extensions and optimisations to the existing protocols for mobility support.

One can assume that in the near future new mobiles will appear that will support multiple wireless interfaces. Therefore, the new version of the Internet Protocol (IPv6) will be one of the next challenges for the wireless community.

For information about other engineering and science titles published by Hermes Penton Science, go to **www.hermespenton.com**